Library of
Davidson College

North American Freshwater Leeches

North American Freshwater Leeches, Exclusive of the Piscicolidae, with a Key to All Species

ROY T. SAWYER

ILLINOIS BIOLOGICAL MONOGRAPHS **46**

UNIVERSITY OF ILLINOIS PRESS   URBANA, CHICAGO, AND LONDON

Board of Editors: Donald F. Hoffmeister, Willard W. Payne, Tom L. Phillips, Richard B. Selander, and Philip W. Smith.

Issued January, 1972.

© 1972 by The Board of Trustees of the University of Illinois. Manufactured in the United States of America. Library of Congress Catalog Card No. 78-166474.

ISBN 0-252-00214-8

ACKNOWLEDGMENTS

Thanks are due to Professor E. W. Knight-Jones, University College of Swansea, Wales; Dr. M. C. Meyer, University of Maine; and Dr. L. R. Richardson, Grafton, New South Wales, for their interest, helpful advice, and criticism. Sincere thanks also go to Drs. I. J. Cantrall, H. van der Schalie, F. C. Evans, and T. E. Moore, University of Michigan, for their encouragement of my initial interest in North American leeches.

Drs. John D. Unzicker and Philip W. Smith, Illinois Natural History Survey, allowed me to examine the large Illinois collection on which much of this work is based. In addition, the following have helped by supplying specimens or locality reports: Dr. R. W. Sims, British Museum (Natural History); Dr. G. Hartwich, Berlin Museum; Dr. Herbert Levi, Museum of Comparative Zoology, Harvard University; Dr. Walter J. Harman, Louisiana State University; Mr. David Foley, University of Michigan; Mr. Allen J. Smith, Hammond Bay (Michigan) Biological Station; and Mr. John G. Hale, Department of Conservation, Duluth, Minnesota.

This study was supported in part by a National Science Foundation Predoctoral Fellowship.

CONTENTS

INTRODUCTION	1
SYSTEMATIC ACCOUNTS	4
ZOOGEOGRAPHICAL AND EVOLUTIONARY CONSIDERATIONS	73
KEY	76
REFERENCES	91
FIGURES	107
INDEX	149

INTRODUCTION

Leeches, like snakes and spiders, fall into the category of "misunderstood" creatures. The popular image of the leech — a vampire-like worm which preys on blood at every opportunity — has arisen from tales of the once-common practice of "leeching" to rid the body of "bad blood" or from accounts of attacks on swimmers by bloodsucking leeches. The bloodsucking activity of a few species is insufficient reason to fear all leeches; the majority of species are restricted to feeding on certain groups of animals, such as fish, salamanders, turtles, birds, and snails.

Leeches are annelid worms, easily recognized by their flattened bodies and sucking discs. Like their relatives, the earthworms, they are hermaphroditic, and the eggs of most species are deposited in cocoons secreted by the leech's body. There is no larval stage, the eggs hatching into miniature leeches. Some leeches migrate upstream en masse in the spring, brood their offspring until almost grown, eat each other's eggs, and parasitize other members of their own group. Some crawl on land at night in search of food; some can predict the weather by responding to changes in barometric pressure; and some can swim, burrow, or crawl with equal ease. Leeches live in fresh water, in the sea, and even on land, and they occur from the polar seas to the tropical jungles. Leeches are of medical importance, for they serve as intermediate or final hosts of several parasitic protozoans, trematodes, cestodes, nematodes, and nematomorphs.

The leeches constitute a significant part of the North American freshwater fauna both in numbers of species and in biological importance, but to date they have received little attention, primarily because of the difficulties of identification. Except for the excellent studies of the family Piscicolidae by Meyer (1940, 1946a), there has been no critical review of North American freshwater leeches since Verrill's (1874a) summary. Since then a few works dealing with morphology (Castle, 1900a; Moore, 1901, 1912) and a few regional studies, primarily from the northern United States and Canada, have been published. Apart from the recent thorough study of Colorado leeches by Herrmann (1970), the Great Lakes region has been the most intensively studied section of North America, as a result of investigations in Minnesota (Moore, 1912), Wisconsin (Bere, 1931; Sapkarev, 1968), Michigan (Miller, 1937; Sawyer, 1968, 1970a), Illinois (Moore, 1901), Indiana (Moore, 1920), Ohio (Moore, 1906; Miller, 1929), and Ontario (Ryerson, 1915; Moore, 1924b, 1936; Meyer, 1937b; Meyer and Moore, 1954; Thomas, 1966).

This monograph is an attempt to revise critically the species of the families Glossiphoniidae, Erpobdellidae, and Hirudinidae, the Piscicolidae having been treated fully by Meyer (1940, 1946a) and Hoffman (1967). This revision entailed examination of 5,000 specimens from over 200 collecting stations in 20 states, although most of the material was from Michigan and Illinois. Hundreds of specimens were examined alive, scores of erpobdellids and hirudinids were dissected, and numerous serial sections and whole mounts were made of the glossiphoniids. Excluding the piscicolids, 40 described species were encountered, plus two new species, and several forms of uncertain taxonomic status were studied.

For each species a complete American synonymy has been compiled. The order of presentation in the text follows the chronological order of the original description of the genera and species within each family and genus respectively. In the systematic accounts emphasis has been placed on identification and on a biological as well as morphological definition of each species. Information is also presented on relative abundance, distribution, individual and geographic variation, and biology. To facilitate identification, the species and the more common variants have been illustrated, and the key has employed as many characters as possible. A full bibliography of the primary literature on leeches of the United States and Canada, including references to the piscicolids, is presented to encourage and aid future investigators.

Although much work is urgently needed on the internal morphology of the glossiphoniids, almost all American species in this family can be accurately identified by numerous external characters alone. Therefore, for the benefit of the nonspecialist the emphasis in the space allotted to the Glossiphoniidae has been placed on a critical analysis of the major variations in pigmentation, papillation, the ocelli, and other important external characters with which the nonspecialist can quickly become acquainted. Unfortunately, an understanding of the internal morphology is necessary for the accurate identification of some of the erpobdellids and hirudinids, and these families are treated accordingly.

Most of the specimens examined are deposited in my personal collection and that of the Illinois Natural History Survey; some specimens from other sources are so indicated in the text. While most of the localities are from Illinois and Michigan, material has also been studied from Arkansas, Colorado, Delaware, Florida, Indiana, Kansas, Kentucky, Maine, Maryland, Minnesota, Missouri, Montana, New York, Pennsylvania, South Carolina, Tennessee, Texas, and Wisconsin. In the interest of saving space, locality data to document the records on the distribution maps are not given here but can be found in Sawyer (1969).

With the exception of Fig. 37, all illustrations were drawn by the author.

SYSTEMATIC ACCOUNTS

Family Glossiphoniidae

GENUS *GLOSSIPHONIA* JOHNSON, 1816

The almost cosmopolitan genus *Glossiphonia* is represented in North America by two widely distributed species, *Glossiphonia complanata* and *G. heteroclita*, both of which also occur throughout most of Eurasia. The genus under the names *Glossiphonia* and *Glossosiphonia* has been a catchall for many unrelated groups which were eventually raised to generic rank: *Helobdella, Placobdella, Batracobdella,* and *Theromyzon*. A key and catalogue to species and subspecies of *Glossiphonia* can be found in Soós (1966c).

For many years the American species of *Glossiphonia* (*"Clepsine" elegans* Verrill, 1872, and *"C." pallida* Verrill, 1872) were thought to be distinct from the European *Glossiphonia complanata* and *G. heteroclita* respectively, but such workers as Castle (1900a) and Moore (1901) concluded that Verrill's nominal species were the same as their European counterparts. The subspecies *Glossiphonia complanata mollissima* Moore and Meyer, 1951, from Alaska probably represents only a color variant.

Glossiphonia complanata (Linnaeus, 1758)

[A full European synonymy can be found in Harding (1910) and Autrum (1936).]

Hirudo complanata Linnaeus, 1758:650.
Clepsine mollissima: Grube, 1871:87; Moore, 1898:547.
Clepsine elegans: Verrill, 1872b:132, fig. 3B; Verrill, 1873b:387; Verrill, 1874a:684; Verrill, 1875a:967; Forbes, 1893:218; Moore, 1898:548; Moore, 1952:4.
Clepsine patelliformis: Nicholson, 1873:493.
Clepsine pallida var. *b:* Verrill, 1874a:684; Verrill, 1874b:623; Moore, 1952:4.
Clepsine pallida: Verrill, 1875a:966.
Clepsine sex-puncto-lineata: Sager, 1878:73.
Clepsine complanata: Graf, 1899:224.
Glossiphonia elegans: Castle, 1900a:46, figs. 5, 11, 28-31.
Glossiphonia complanata: Moore, 1901:493; Moore, 1906:156, fig. 4; Moore, 1912:82, fig. 4; Ryerson, 1915:165; Hankinson, 1916:118; Moore, 1918:652; Moore, 1920:89; Moore, 1922:7; Moore, 1923:15, pl. 1B; Moore, 1924b:21; Mullin, 1926a:35, pl. VI, figs. 4-5; Bere, 1929:177; Miller, 1929:10, fig. 2; Bere, 1931:438; Moore, 1936:113; Meyer, 1937a:249; Meyer, 1937b:118; Miller, 1937:85; Richardson, 1942:68; Richardson, 1943:89; Mathers, 1948:397, pls. 1, 4; Pawlowski, 1948:329; Moore and Meyer, 1951:58; Moore, 1952:4; Pennak, 1953:315, fig. 200H; Beck, 1954:74; Meyer and Moore, 1954:67; Oliver, 1958:163; Moore, 1959:548; Paloumpis and Starrett, 1960:416; Mann, 1961b:157; Moore, 1964:1; Moore, 1966a:10; Thomas, 1966:202; Sawyer, 1967:36; Carlson, 1968:164; Sapkarev, 1968:226; Sawyer, 1968:228; Scudder and Mann, 1968:208; Clifford, 1969:583; Herrmann, 1970:5.
Glossiphonia complanata mollissima: Moore and Meyer, 1951:59.
Glossophiona complanata: Mason et al., 1970:R323.

Description (Fig. 1A). This species is easily recognized by the three pairs of eyes (Fig. 17A) and by the distinctive dorsal and ventral pigmentation. It might be confused with *Glossiphonia heteroclita*, *Placobdella hollensis*, and *Theromyzon* sp. However, the three pairs of eyes in *Glossiphonia complanata* are arranged in two longitudinal rows, and, unlike *G. heteroclita*, eyes of the first pair are not ordinarily closer together than the other two. *Theromyzon* has four pairs of eyes, the first pair of which can easily be overlooked, whereas *Placobdella hollensis* has actually one pair of eyes, the two apparent eyes behind them being metameric pigment concentrations.

A characteristic pair of strong but narrow dark paramedial stripes extends from the anal region to the anterior neck region, where they approach each other, and then diverge slightly as they proceed anteriorly lateral to the eyes. These stripes and similar ones on the ventral surface usually remain after preservation, even after other pigments have faded. Two pairs of metameric paramedial white dots, along with a few other scattered dots, extend from the anal region to the anterior third of the body. As in *Placobdella* and *Batracobdella*, but not in the allied *Glossiphonia heteroclita*, there is a marginal metameric series of dots. Unlike *G. heteroclita*, which is generally translucent and unpigmented, *G. complanata* is almost always an opaque brown with white dots and blackish stripes.

Variation. There is slight variation in the size and position of the eyes, especially the first pair, which may be smaller than the others and occasionally closer together, approaching the condition in *G. heteroclita*. In some individuals the dorsal paramedial stripes are more or less continuous, but usually they are interrupted metamerically to varying degrees by the paramedial pairs of white metameric dots. The stripes, which are most strongly developed in the anterior third of the body, may be faded or missing posteriorly. The ventral stripes are usually continuous but are sometimes irregularly interrupted. The dorsal dots also vary from small distinct spots to large diffused patches which may fuse together. The centers of these dots are sometimes slightly raised, rarely to such an extent that they appear to be small metameric papillae. The two paramedial rows of dots resemble those found in *Theromyzon* sp., *Batracobdella picta*, and *Helobdella lineata*, all of which differ from *Glossiphonia complanata* in the number of eyes.

The degree of pigmentation varies considerably from one locality to another. At one extreme, primarily in leafy or muddy situations, are populations so darkly pigmented that the paramedial stripes may be almost obliterated by the densely packed, blackish chromatophores. In such dark individuals the two rows of paramedial dots are almost always present, often tending to be yellowish rather than white. At the other extreme, usually along sandy beaches or limestone rocks, are light-colored populations in which the patterns of white dots may vary from the typical condition described above to large, indistinctly fused white patches, obscuring the basic pattern to become an ornate splotchy design or, rarely, an almost uniform near white. However, even in such lightly pigmented individuals the anterior portions of the paramedial stripes can usually be seen. Rarely can an individual be found in which the paramedial stripes are completely missing both dorsally and ventrally, and even then the body is opaque rather than translucent as in *G. heteroclita*.

Ecology. Various aspects of the ecology and natural history of British *G. complanata* were examined by Mann (1955, 1956, 1957b), who found the species in almost every type of freshwater habitat. In North America the species feeds predominantly on snails. Moore (1964) found that in the laboratory it would feed on *Physa hetero-*

stropha and *Menetus exacuous* but not on *Lymnaea emarginata*. In Michigan I observed it feeding on a *Physa* while still carrying young, which, if judged from their filled caeca, had also fed. Thus the young do not necessarily leave the parent with the first feeding. Individuals of *G. complanata* were also found on adult *Haemopis grandis* and *Erpobdella punctata*, especially in quiet mud-bottomed ponds and streams. Its ecology is similar in Europe (Jarry, 1960).

In Michigan, as in England (Mann, 1957b), insemination is by hypodermic implantation of spermatophores, which are often found on the ventral surface. Brooding individuals have been encountered by myself and others on the following dates — 5, 19, and 30 April; 8, 18, and 25 May; and 5 and 6 June — suggesting that breeding occurs primarily in spring. In Michigan most of the individuals laid their eggs around 30 April, when the water temperature was about 15°C. Unlike *Helobdella*, *Batracobdella*, and *Oculobdella*, the cocoons are attached to the substrate, not to the ventral surface of the parent. The eggs are tightly enveloped in a delicate membranous sac containing little albumen. On the average each individual produced 6.24 cocoons, each containing 20.6 eggs (Figs. 6A, B). Mann (1957b) found that in England a mean of 33 eggs per individual was brooded, in marked contrast to the situation in Michigan, where the mean was about 129 eggs. When forcibly removed, the brooding leeches usually returned to their own cocoons, but on two occasions they covered cocoons belonging to another individual.

Distribution (Fig. 19). *Glossiphonia complanata* is one of the most common leeches in North America. It occurs in most parts of Eurasia, Canada, and the United States, except possibly in the poorly studied extreme western and southern states. Published records believed valid are from Alaska, Northwest Territories (Moore and Meyer, 1951), the southern tip of the Georgian Bay (Ryerson, 1915), British Columbia (Clemens *et al.*, 1939; Meyer and Moore, 1954; Scudder and Mann, 1968), Alberta (Bere, 1929; Moore, 1964; Clifford, 1969), Saskatchewan (Moore and Meyer, 1951; Oliver, 1958), Ontario (Moore, 1906; Faull, 1913; Moore, 1924b, 1936; Meyer and Moore, 1954; Thomas, 1966), Quebec (Meyer and Moore, 1954), Newfoundland, Nova Scotia (Pawlowski, 1948), New Brunswick, Prince Edward Island (Richardson, 1943), Oregon (Mason *et al.*, 1970), Utah (Beck, 1954), Colorado (Verrill, 1874b; Herrmann, 1970), Iowa (Mathers, 1948; Carlson, 1968), Minnesota (Moore, 1912), Wisconsin (Verrill, 1875a; Bere, 1931; Sapkarev, 1968), Michigan (Adams, 1908; Hankinson, 1916; Miller, 1937; Sawyer, 1968), Illinois (Moore, 1901; Paloumpis and Starrett, 1960), Indiana (Moore, 1920), Ohio (Moore, 1906; Miller,

1929), Pennsylvania (Moore, 1912), New York (Moore, 1923; Barrow, 1953), Connecticut (Verrill, 1874b; Barrow, 1953), and Massachusetts (Castle, 1900a). It is known as far south as extreme southeastern Missouri (Meyer, 1937a).

Glossiphonia heteroclita (Linnaeus, 1761)

[A full European synonymy can be found in Harding (1910) and Autrum (1936).]

Hirudo heteroclita Linnaeus, 1761:364.
?*Clepsine swampina:* Diesing, 1850:435; Verrill, 1872b:132; Verrill, 1874a: 685; Moore, 1952:4.
Clepsine pallida: Verrill, 1872b:131, fig. 3; Verrill, 1874a:684, fig. 2; Verrill, 1874b:623; Moore, 1952:4.
Glossiphonia heteroclita: Castle, 1900a:42, figs. 19-22, 35-36, 38; Moore, 1906:156; Ryerson, 1915:165; Moore, 1918:652; Moore, 1920:89; Moore, 1922:7; Mullin, 1926a:35; Bere, 1931:438; Mathers, 1948:397, pls. 1, 4; Pawlowski, 1948:330; Kenk, 1949:38; Moore, 1952:4; Pennak, 1953:315, fig. 200G; Meyer and Moore, 1954:67; Moore, 1959:548; Mann, 1961b: 157; Sapkarev, 1968:226; Sawyer, 1968:228.

Description (Fig. 1B). This species is characterized by three pairs of eyes (Fig. 17B), the first pair of which is closer than the two posterior pairs, and by a gelatinous-like translucent body in which is seen the conspicuous internal organs, especially the crop. It differs from *G. complanata* in that the eyes of the latter are equidistant in two longitudinal rows, and the body is invariably opaque with characteristic patterns on the dorsal and ventral surfaces. *Glossiphonia heteroclita* may easily be mistaken at first glance for the more common *Helobdella stagnalis* or *H. elongata*, but it differs from both in the number of eyes and in having a less elongate body.

Variation. Generally the body is whitish and devoid of pigment, but some individuals have inconspicuous, fine black chromatophores in sparse clumps of from one to six. They are situated metamerically from the anal region to the head, especially along the middorsal line and along the margins of the posterior part of the body. In other individuals numerous brownish gland cells can be seen through the body wall. The eyes of the first pair vary somewhat in size and relative positions. It is common for one of the eyes to be minute or missing, and both eyes of the first pair may be missing in an occasional individual.

Ecology. The biology of this species was studied in Wales by Hatto (1968) and Gruffydd (1965). The latter showed that the species inhabits the mantle cavity of the snail *Lymnaea pereger* from October to May, with a peak infestation around January. During the breeding season, from May to October, the leeches are free-living. He concluded that the life histories of *G. heteroclita* and *Lymnaea pereger* are intimately related, but there is as yet no evidence of such a relationship between *G. heteroclita* and a snail in North America.

On one occasion I found spermatophores attached to the ventral surface in the genital region. Individuals I collected on 21 May and 13 July were obviously gravid, suggesting that in the midwestern United States, as in Great Britain, breeding occurs in the summer. Castle (1900a) reported that between 11 and 65 eggs are deposited, depending upon the size of the individual. Various authors (Castle, 1900a; Moore, 1920) have observed that, unlike those of other known glossiphoniids, the eggs of the species are attached singly (rather than in membranous capsules) to the ventral surface of the parent. Hatto (1968), on the other hand, reports that the eggs are in capsules.

Distribution (Fig. 20). *Glossiphonia heteroclita*, known from Eurasia and North America, is represented in the present study by a few individuals from southern Michigan and northeastern Illinois. Published records believed reliable include the southern tip of the Georgian Bay (Ryerson, 1915), Alberta (Moore, 1964), Manitoba (Meyer and Moore, 1954), Ontario (Moore, 1906; Faull, 1913), Newfoundland (Pawlowski, 1948), Iowa (Mathers, 1948), Wisconsin (Bere, 1931; Sapkarev, 1968), Michigan (Kenk, 1949; Sawyer, 1968), Indiana (Moore, 1920), Pennsylvania (Moore, 1906), Connecticut (Verrill, 1874a), and Massachusetts (Castle, 1900a). It is known at present only as far south as northern Indiana (Moore, 1920), which may merely reflect difficulties in finding and identifying the species. It has not been previously recorded from Illinois.

GENUS *BATRACOBDELLA* VIGUIER, 1879
(NOT *BATRACHOBDELLA* CABALLERO, 1931)

The several species of *Batracobdella* are among the least known of all the North American leeches. They are small, relatively scarce, and inadequately described. Three species have previously been reported from North America: *B. paludosa* (Carena, 1824), *B. picta* (Verrill, 1872b), and *B. phalera* (Graf, 1899). The record for *B. paludosa*, a European leech characterized by two pairs of eyes, a feature not otherwise found in North American *Batracobdella*, is based on one specimen from Newfoundland (Pawlowski, 1948) and, to my knowledge, has never been confirmed there or elsewhere. Unless more specimens are found, it is best to consider the species as not established in North America; the other two species are commonly encountered. Their true systematic positions are still unsettled. For example, Soós (1967) in his excellent review of the genus erroneously reassigned the two species to *Placobdella* because of the lack of adequate descriptions. In addition, a new species of *Batracobdella* encountered in the present study is described herein.

Batracobdella picta (Verrill, 1872)

Clepsine picta Verrill, 1872b:128; Verrill, 1874a:678; Verrill, 1875a:965; Moore, 1952:3.
Placobdella picta: Moore, 1906:157, fig. 3; Ryerson, 1915:166; Moore, 1918: 653; Miller, 1929:10; Bere, 1931:439; Moore, 1936:113; Meyer, 1937a:250; Miller, 1937:85; Kenk, 1949:38; Pennak, 1953:315; Soós, 1967:243.
Glossiphonia picta: Moore, 1923:15.
Haementeria? (Placobdella) picta: Autrum, 1936:78.
Batrachobdella picta: Richardson, 1949:85; Moore, 1952:3; Barrow, 1953: 197; Beck, 1954:74; Moore, 1959:548, fig. 23.2.
Batracobdella picta: Meyer and Moore, 1954:67; Mann, 1961b:158; Sapkarev, 1968:226; Sawyer, 1968:228; Brockleman, 1969:632; Herrmann, 1970:5.

Description (Fig. 2C). *Batracobdella picta* is characterized by a smooth dorsal surface, a single confluent pair of eyes, four series of white dots, and often a dark middorsal stripe. Usually there are about 15 longitudinal rows of dark chromatophores, which are especially conspicuous along the margins and along the middorsal line, producing the characteristic middorsal stripe. On preservation, the white metameric dots and, to a lesser extent, the middorsal stripe may become obscure. The eyes of *B. picta*, usually surrounded by a white unpigmented area, are typically confluent but are separated by a short distance in occasional specimens. In many specimens a variable white transverse stripe or ring is present in the neck region. This ring may be completely missing or from one to three annuli wide, when it is as conspicuous as that of *B. phalera* or *B. michiganensis*. Earlier keys erroneously separated *B. phalera* from *B. picta* by the white area surrounding the eyes and by the presence of the ring. *Batrocobdella phalera* and *B. michiganensis* are both distinguished from *B. picta* by having a translucent body with a short conspicuous white bar above the genital region. Some specimens of *B. picta* may have a poorly developed middorsal row of dots, and under ideal lighting conditions metameric white patches can be discerned along the lateral margins, a feature distinctive in other species of *Batracobdella* and *Placobdella*.

Remarks. A full description of this species promised by J. Percy Moore (Meyer and Moore, **1954:67**) was never published, nor were any notes on the species found among his papers (Meyer, *in litt.*).

Ecology. This species, which is usually found in small woodland ponds, is one of the earliest to appear in the spring. As early as late March, when ice was still on most of the ponds in Michigan, they could be found sluggishly active along the ice-free edges, where the water temperature was 3.5°C. Their early appearance and their habit of aggregating in the shallow warmer water along the edges of the ponds may be related to the arrival and breeding habits of their amphibian hosts. They are often extremely abundant locally and can be found attached to most, if not all, species of amphibians which frequent the ponds. They seem to feed exclusively on amphibians but are not otherwise host-specific. I observed this species feeding on adult *Ambystoma tigrinum,* adult and larvel *Bufo americanus* and *Rana catesbeiana,* and larval *Hyla versicolor* and *H. crucifer.*

The importance of *B. picta* in regulating natural populations of amphibians was thoroughly studied by Brockleman (1968, 1969), who found that the number of young *B. picta* can be as high as 66 per square meter; under such high density predation by *B. picta* is the largest single source of mortality of the tadpoles of *Bufo americanus.* Under seminatural conditions he found that mortality from leech predation was highest where density of tadpoles was high, rather than low, which suggests that leeches are differentially attracted to situations of high tadpole density. The young are found on tadpoles of moderate to large size, usually attached at the base of the tail, where they interfere least with swimming. On one occasion I found a young *B. picta* feeding on a recently hatched tadpole of *Hyla versicolor* only 10 mm long. Even on larger tadpoles they can often kill their hosts within one to two days.

Barrow (1953) showed that *B. picta* can play a major role in the transmission of *Trypanosoma diemyctyli* (Tobey) into the newt *Notophthalmus viridescens.* The newt becomes thoroughly infected with trypanosomes 12 to 16 days after initial infection, and the cycle is perpetuated when the young of *B. picta* feed on infected adult newts.

Barrow's observation that feeding by the adult leeches induces them to breed agrees with my field observations. Mating occurs very early in the spring, in Michigan as early as the first week in April, when the water was 5.6°C. In every case when mating was observed, the spermatophores were attached ventrally in the region of the genital openings. The young were found throughout the summer months on various larval amphibians, especially *Bufo americanus,* and there is every indication that the young breed early the next spring.

Batracobdella picta has been found on rare occasions in the dorsal subcutaneous lymph spaces of *Rana catesbiana* (Richardson, 1949). The regulation of population density, the transmission of trypanosomes, and the occurrence of endoparasitism indicate that *B. picta* has developed an intricate relationship with amphibians of long evolutionary standing.

Distribution (Fig. 20). *Batracobdella picta* is a widely distributed species which is relatively uncommon but often locally abundant. It is known from Colorado (Herrmann, 1970), Utah (Beck, 1954), southeastern Missouri (Meyer, 1937a), Wisconsin (Bere, 1931; Sapkarev, 1968), southern Georgian Bay (Ryerson, 1915), Ontario (Faull, 1913; Moore, 1936), Quebec (Richardson, 1949; Meyer and Moore, 1954), New York (Moore, 1923; Barrow, 1953), Connecticut (Verrill, 1874a; Barrow, 1953), Michigan (Miller, 1937; Kenk, 1949; Sawyer, 1968), and Ohio (Moore, 1906; Miller, 1929). It has not been previously recorded from Illinois.

Batracobdella phalera (Graf, 1899)

Clepsine phalera Graf, 1899:354, figs. 116-118.
Placobdella phalera: Moore, 1906:157; Ryerson, 1915:166; Moore, 1918:654; Moore, 1922:7; Moore, 1923:15; Baker, 1924:109; Mullin, 1926a:55; Miller, 1929:10; Bere, 1931:437; Miller, 1937:90; Richardson, 1942:68; Richardson, 1943:90; Mathers, 1948:397, fig. 2; Pawlowski, 1948:318; Pennak, 1953:315; Meyer and Moore, 1954:84.
Placobdella phaleria: Mullin, 1926a:36.
Haementeria (Placobdella) phalera: Autrum, 1936:77, fig. 50.
Batrachobdella phalera: Moore, 1959:548.
Batracobdella phalera: Mann, 1961b:158; Sapkarev, 1968:226; Herrmann, 1970:5.

Description (Figs. 2D, E). *Batracobdella phalera* is characterized by three series of dark dorsal papillae, a convex translucent body, a white bar on the dorsum above the genital area (sometimes with another bar about two-thirds of the way caudad), and a conspicuous white anal patch. The body is sometimes flattened, and there may be a darkly pigmented stripe along the middorsal row of papillae and darkly pigmented metameric markings along the margins. It is distinguished from *B. picta* by the white metameric markings along the margins, the dorsal papillae, the convex translucent body, and the absence of the four series of metameric dots.

Remarks. A poorly known species, *B. phalera* was originally described from Falmouth, Massachusetts (Graf, 1899), reportedly parasitic on the common musk turtle (*Sternothaerus odoratus*). But parasitism of turtles is a characteristic of *Placobdella*, not *Batracobdella*, the amphibian leeches, which suggests that Graf may have described a species of *Placobdella*. To my knowledge the leech as described has never been reported since, but a superficially similar leech from western

Lake Erie has become somewhat doubtfully accepted as the name-bearer of *B. phalera* (see Moore, 1906). The species found in the present study closely resembles Moore's species but differs in having few or no papillae along the margin of the hind sucker. Most specimens examined had three dorsal rows of dark-tipped papillae, but a few had unpigmented paramedial rows of papillae, appearing as white dots. In addition, there was a much more conspicuous dark brown mid-dorsal stripe from segments VIII to XII (Fig. 2E).

Ecology. Whereas *B. picta* inhabits small woodland ponds, *B. phalera* is found only along the edges of much larger bodies of water such as lakes and rivers. Except for questionable reports of the occurrence of *B. phalera* on a turtle (Graf, 1899) and on a bluegill (Bere, 1931), the hosts of this species are unknown. The occurrence of brooding individuals on 6 June and 22 July, when the water temperature was 21-22.5°C, suggests that this species breeds in mid-summer, unlike *B. picta*. Summer breeding in *B. phalera* is further corroborated by the observations of Moore (1922) and Mathers (1948). On 15 August in Nova Scotia the former author found a brooding individual with four cocoons, each containing four to six eggs.

Distribution (Fig. 21). *Batracobdella phalera* appears to be widely distributed and reasonably abundant around the Great Lakes. Published reports believed valid are from Manitoba (Meyer and Moore, 1954), the southern tip of the Georgian Bay (Ryerson, 1915), Ontario (Moore, 1906; Faull, 1913), Nova Scotia (Pawlowski, 1948), Colorado (Herrmann, 1970), New York (Moore, 1923), Iowa (Mathers, 1948), Wisconsin (Baker, 1924; Sapkarev, 1968), and Ohio (Moore, 1906; Miller, 1929). The discovery of specimens of this species from Illinois and Michigan in the present study represents the first records for those states.

Batracobdella michiganensis, new species

Type-Locality. St. Joseph County, Michigan (Mill Creek at Young's Prairie Road, 3.5 miles south of Michigan State Highway 60). Types deposited in Charleston (S.C.) Museum. Holotype: 71.20.1. Paratype: 71.20.2.

Description. On 28 May 1967 eight specimens of this small undescribed species, only the diagnostic characters of which will be presented here, were found in St. Joseph County, Michigan. They are characterized by being excessively flattened, shaped like *Salix* leaves, rather pale and translucent (Fig. 2A). The eyes are fused in some individuals, barely touching in others. In addition to the white margins, each has five distinct longitudinal rows of white prominences, surrounded by yellowish dots equidistant from each other longitudinally and transversely, and many fine longitudinal and transverse stripes, giving a uniformly checkered appearance. There are four white patches dorsally, one less than in most *B. phalera*, the first around the eyes, the second forming a ring around the neck (in such a way as to isolate a darkly pigmented band behind the eyes), the third in the general region of the clitellum, and the fourth in the anal region. The ventral surface is generally unpigmented, except for fine longitudinal stripes and the dorsal yellow dots, which are visible through the translucent body. These leeches were about 6 mm long, with a circular hind sucker 0.75 mm in diameter. Judged from the numerous well-developed eggs in the ovisacs, they were probably sexually mature. Complete transverse serial sections of the paratype, cut at 10 μ and stained with eosin and Ehrlich's haematoxylin, revealed six pairs of testes, the first pair being displaced somewhat anteriorly. Reconstructions of the digestive and reproductive systems of the types are substituted for a lengthy verbal description of these systems, which are typical of the genus (Fig. 2B).

Unlike *B. phalera* and *B. picta, B. michiganensis* has five series of slightly raised metameric dots, including a middorsal series. It differs from *B. picta* in its smaller size, the absence of a middorsal stripe, the presence of white patches in the clitellar, genital, and anal regions, and the presence of a middorsal series of metameric dots and metameric markings along the margins. It differs from the various species of *Placobdella* by the distinctive metameric markings and by the absence of distinct papillae, although it has slightly raised metameric dots which could be taken for papillae. It differs from *Placobdella papillifera*, with which it was found, in having a translucent body, confluent eyes, and white patches in the neck, clitellar, and anal regions.

GENUS *THEROMYZON* PHILIPPI, 1884

There is considerable confusion concerning the systematics of the various North American species of *Theromyzon*, the true bird leeches. The earliest record is a vague description by Baird (1869) of *Glossiphonia rudis* from Great Bear Lake. *Glossiphonia rudis* was known only from the original description until it was rediscovered, again from Great Bear Lake, by Moore and Meyer (1951), who showed that the gonopores were separated by three annuli. A second species, *Clepsine occidentalis*, which was described from Colorado by Verrill in 1874, was subsequently characterized by having only two annuli between the gonopores (Moore, 1912, 1918; Pennak, 1953). For almost

70 years almost all North American *Theromyzon* were identified as *T. occidentalis* until the rediscovery of *T. rude* strongly suggested that Verrill's *occidentalis* was actually identical with *T. rude*, a species apparently characteristic of the Rocky Mountain area. The form with two annuli, which appears to be characteristic of the Mississippi River Basin, has been given the same name as a similar European leech, *T. meyeri* (Livanow, 1902). A third species, *T. tessulatum*, which is distinguished by the gonopores being separated by four annuli, is characteristic of Eurasia but has been reported from North America (Pawlowski, 1948; Oliver, 1958; Herrmann, 1970).

The presence of three distinct species in North America is assumed by Oliver (1958), who reported without comment all three species from Saskatchewan. Considering the ease with which these leeches can be transported by birds, it is more likely that they are all variants of the same species, which may or may not be the same species as the European *T. tessulatum*. The validity of separating *T. rude* and *T. meyeri* solely on the basis of the number of annuli between the gonopores is open to serious question, especially after Meyer and Moore (1954:85) reported that the gonopore separation is subject to some variation. Until this problem is critically examined, it is best to treat separately the two distinctly American forms, *T. rude* and *T. meyeri*.

Theromyzon rude (Baird, 1869)

Glossiphonia rudis Baird, 1869:317; Autrum, 1936:46.
?*Clepsine occidentalis:* Verrill, 1874a:685; Verrill, 1875a:966; Moore, 1952:4.
Theromyzon occidentalis: Bere, 1929:177; Fredeen and Shemanchuk, 1960:733.
Theromyzon rude: Moore and Meyer, 1951:60; Moore, 1952:4; Meyer and
    Moore, 1954:84, pl. 1, fig. 3; Oliver, 1958:163; Moore, 1959:548, fig. 23.1;
    Mann, 1961b:155; Moore, 1964:1; Hagadorn, 1966a:288; Moore, 1966a:10;
    Scudder and Mann, 1968:208; Herrmann, 1970:5.

Remarks. I recently found Baird's type in the British Museum (Natural History) with the following label: "*Glossiphonia rudis* Baird, 1869, Type, 1849:10:29:1, Loc: Great Bear Lake, N. America, Pres: Sir J. Richardson, M.D., Ref: Proc. Zool. Soc. London, 1869, p. 317." In the vial were two specimens, apparently belonging to different species. One specimen closely resembles *Placobdella ornata*, with each annulus covered with numerous papillae, a middorsal row, and a paired paramedial row of papillae being somewhat larger than the others. Only one pair of eyes could be seen, and the gonopore separation was obscured.

In the second specimen only three pairs of eyes could be seen, the anterior pair being closer than the other two pairs. The specimen had faded, and the anterior tip of the head was folded into the sucker cavity; if another pair of eyes existed, it may well have been obscured. The dorsal surface had two pairs of paramedial dots or prominences, in addition to some smaller prominences, on the neural annulus.

The presence of two specimens belonging to different species in the vial strongly suggests the possibility that Baird based his brief description on both specimens. Two points in the original description corroborate this: "roughly annulated, each ring armed with a series of tubercles along its surface" applies only to the *Placobdella ornata*-like specimen, whereas "eyes ? six in number (as far as could be made out)" applies only to the *Theromyzon*–like specimen.

The occurrence in the midwestern United States of a *Theromyzon* with three annuli between the gonopores has never been substantiated, but such a form, which may or may not be distinct from *T. meyeri*, may eventually be found. Although the known distribution of *T. rude* is sketchy, it appears to be characteristic of the Rocky Mountain and Great Basin regions (Herrmann, 1970), possibly associated with well-established bird migration routes. Detailed examinations of neurosecretion and its role in reproduction have been made on this species (Hagadorn, 1958, 1961, 1962, 1966; Hagadorn *et al.*, 1963). Certain neurosecretory cells in the brain of *T. rude* have an annual cycle of secretion, correlating with spermatogenesis and showing a peak in the spring and early summer months. Other aspects of the natural history of this species, which is known to infest many species of water birds, have been investigated by Meyer and Moore (1954) and Moore (1964, 1966a), but an exhaustive study is lacking. There is little doubt that *T. rude*, like its European congenitor, *T. tessulatum*, can be a cause of morbidity and mortality of young waterfowl, but the economic significance of this problem requires critical examination.

Theromyzon meyeri (Livanow, 1902)

Protoclepsis meyeri Livanow, 1902:339.
Hemiclepsis occidentalis: Moore, 1912:96, fig. 12; ?Oliver, 1958:163; ?Moore, 1964:8.
Protoclepsis occidentalis: Moore, 1918:654, fig. 999; ?Moore, 1922:7.
Theromyzon occidentale: Autrum, 1936:45; ?Mathers, 1948:397, pls. 2, 4; Pennak, 1953:315, fig. 200J; ?Meyer and Moore, 1954:66; Oliver, 1958:163.
Theromyzon occidentalis: Sooter, 1937:108; Richardson, 1943:89; Richardson, 1949:85; ?Moore, 1964:8.
Theromyzon meyeri: Moore, 1959:548, fig. 23.1; Mann, 1961b:155; Moore, 1964:1; ?Moore, 1966a:10; Sawyer, 1968:228.

Description (Fig. 1C). In spite of some variation in the body shape and dorsal pigment pattern, this species is immediately distinguishable by the four pairs of eyes (Fig. 17D). Unfed relaxed specimens can often be recognized at a glance by the broad blunt head, which is about the same width as the neck. After engorgement with blood, the usually flattened body can become severely distorted to a globular or sausage shape. The dorsal pigmentation varies in the number and the distinctness of the yellow dots and relative abundance of dark chromatophores. The illustrated specimen from Michigan had only two pairs of small faded dots, although there was a vague hint of a marginal pair, and many conspicuous dark chromatophores were scattered throughout the dorsal and ventral surfaces. On the other hand, another specimen from Michigan had a brilliantly colored dorsal pattern of large yellow dots, each almost the width of an annulus, on a uniform light brown background which contained few dark chromatophores except at the margins and on the ventral surface. In addition to the two pairs of paramedial dots, there was a pair of marginal dots, a less distinct middorsal row, and a circular row near the margins of the hind sucker, as well as a few scattered dots on the dorsum.

Ecology. In contrast to *T. rude*, very little is known about the general biology of *T. meyeri*. Sooter (1937) reported that *T. occidentale* (?*T. meyeri*) from northwestern Iowa infested young waterfowl in July and August so heavily that mortality resulted from obstructions of their air passages. The known hosts for *T. meyeri* are the coot (*Fulica americana*), pied-billed grebe (*Podilymbus podiceps*), and blue-winged teal (*Querquedula discors*) (Sooter, 1937). On 4 June 1967 a Michigan specimen that had been brought into my laboratory several days earlier deposited seven cocoons in a straight row along the side of a glass container. From anterior to posterior each cocoon contained 20, 27, 29, 33, 23, 28, and 20 eggs, or 209 eggs in all, which suggests that *T. meyeri*, like *T. rude*, has a high reproductive potential. Each cocoon was attached individually to the glass substrate by means of a short pedicel and base and resembled those reported for *T. rude* by Meyer and Moore (1954), except that the cocoons were not joined at their bases. The parent was positioned over the cocoons and was not attached to them in any way. Both the oral and caudal suckers were attached to the substrate, while the cocoons were ventilated by periodic downward inflections of the sides of the body, each cocoon bouncing freely on its stalk. Occasionally the inflected sides of the body gripped the cocoons, and the whole body (excluding the caudal sucker) pulled the stalked cocoons somewhat caudad. The ventilatory motion was then resumed with both suckers attached to the substrate. The parent covered the eggs for six days, but the eggs subsequently failed to develop. On 20 June 1967 a brooding individual of *T. meyeri* with large numbers of young attached to its ventral surface was encountered in southwestern Michigan (water 25.5°C).

Distribution (Fig. 21). A species of *Theromyzon*, thought to be *T. meyeri* and represented in the present study from a few localities in Michigan and Illinois, has previously been reported from South Dakota (Moore and Meyer, 1951), Minnesota (Moore, 1912), Iowa (Sooter, 1937; Mathers, 1948), Michigan (Sawyer, 1968), Saskatchewan (Oliver, 1958), and Prince Edward Island (Moore, 1922). Some of these records may be confused with the more western *T. rude*, with which it may prove identical.

GENUS *PLACOBDELLA* BLANCHARD, 1893

Some authors consider that *Haementeria* de Filippi, 1849, has priority over *Placobdella* Blanchard, 1893, but Autrum (1936) established *Placobdella* as a subgenus of the European genus *Haementeria*. There has been such overwhelming acceptance of *Placobdella* as a full genus by North American hirudinologists that the designation will be followed here. *Placobdella* is well represented in North America with seven species; except for *P. pediculata* and *P. montifera*, they feed primarily upon turtles.

Some, if not all, of our *Placobdella* can swim, at least when young. *Placobdella hollensis* is a strong swimmer even as adult, and the young of *P. ornata* and, to a lesser extent, *P. parasitica* can swim for short distances. According to Moore (1912:96), *P. montifera* is also able to swim.

The degree of papillation in the various species of *Placobdella* depends on whether the individual is starved or full when killed and upon the method of preservation. The papillae are more likely to be protruded if the living animal is placed suddenly into preservative without prior relaxation.

Placobdella parasitica (Say, 1824)
Hirudo parasitica: Say, 1824:14; Moore, 1952:8.
Clepsine parasitica: Diesing, 1850:450; Verrill, 1872b:128; Verrill, 1874a: 678; Whitman, 1891:407; Moore, 1952:3.
Clepsine marmorata: Sager, 1878:73.
Clepsine chelydra: Whitman, 1891:418.
Clepsine plana: Whitman, 1891:411, pl. XIV, figs. 1-7, pl. XV, figs. 1-3; Whitman, 1892:392; Bristol, 1898:55; Castle, 1900a:51, figs. 6, 32, 37; Moore, 1952:8.
Glossiphonia parasitica: Moore, 1898:548; Castle, 1900a:51.
Clepsine parasita: Graf, 1899:225.
Clepsine chelydrae: Castle, 1900a:51; Moore, 1952:8.

Glossiphonia parasitica var. *plana:* Castle, 1900a:51.
Placobdella parasitica: Moore, 1901:480, figs. 1, 4; Ward, 1902:278; Moore, 1906:157; Hankinson, 1908:232; Moore, 1912:84, figs. 7, 8; Cahn, 1915: 123; Ryerson, 1915:166; Moore, 1918:653, fig. 998; Evermann and Clark, 1920:304; Moore, 1920:90; Moore, 1922:7; Moore, 1923:15, pl. 1D; Mullin, 1926a:35, pl. IV, fig. 2, pl. V, fig. 3, pl. VI, figs. 1-3; Miller, 1929:10; Bere, 1931:439; Myers, 1935:618, figs. A, 1-18; Moore, 1936:113; Meyer, 1937a:249; Miller, 1937:85; Townes, 1937:167; Richardson, 1942:70; Smith, 1942:410; Mathers, 1948:397, pl. 2; Pawlowski, 1948:329; Kenk, 1949:38; Moore, 1952:3; Moore, 1953:4; Pennak, 1953:315; Meyer and Moore, 1954:84; Oliver, 1958:163; Moore, 1959:550, figs. 23.2, 23.4; Mann, 1961b:159; Moore, 1964:1; Moore, 1966a:10; Patrick *et al.*, 1966:343; Sawyer, 1967:33; Sapkarev, 1968:226; Sawyer, 1968:228; Meyer, 1969:161; Herrmann, 1970:5.
Haementeria (*Placobdella*) *parasitica:* Autrum, 1936:69, fig. 46; Pawlowski, 1948:329.

Description (Fig. 3C). This species is most reliably recognized by the eight to twelve dark longitudinal stripes on the venter, although these may have faded in specimens preserved for some time in alcohol. The dorsum is characterized by great variation of the pigment pattern, probably no two specimens being exactly alike. Basically, there are two main types of dorsal pigmentation, the distinction of which is somewhat arbitrary because intermediate forms do occur. One, similar to that illustrated in Fig. 3C and found in approximately 60 percent of the specimens, is characterized by a broad cream-colored middorsal stripe which bulges laterad in five or six places between the neck and anus. Between this broad stripe and the margins are irregularly shaped cream-colored patches which are more or less metamerically arranged and are often irregularly fused with one another. The second type, found in 30 percent of the individuals, is characterized by a much narrower cream-colored stripe which bulges laterad only slightly, if at all. The regions between this stripe and the margins are larger, more pigmented with brown, and have a distinct row of smaller metameric dots, rarely fusing into one another. The occurrence of neither form could be correlated with geographical distribution, habitat, or time of year collected. Two other pattern types were also encountered on only one or two occasions: one found in northeastern Illinois had a continuous narrow middorsal *brown* stripe flanked on either side by seven or eight smaller darkish longitudinal stripes, much like that typically found on the venter of this species. The other, encountered in western Pennsylvania, had no middorsal stripe at all and was covered dorsally with scattered, irregularly shaped cream-colored patches, whereas the venter had the longitudinal stripes typical of the species.

Most specimens had a smooth dorsum, but some were found with varying degrees of papillation. None, however, even approached the condition found in most *P. ornata*. The dorsum of living *P. parasitica* is a mixture of green, cream-colored, and reddish-brown pigments, but most preserved specimens had only the latter two because the green pigment quickly dissolved in alcohol. The species has the ability to change quickly from a dark green to a dark brown after collection. The mechanism of color change in this species was examined by Smith (1942).

Ecology. Although it has been known to attack other hosts occasionally, such as the tadpoles of *Rana pipiens* (Meyer and Moore, 1954) and fish (Ryerson, 1915; Pearse, 1924), *P. parasitica* feeds predominantly upon turtles, especially the snapping turtle (*Chelydra serpentina*). It is, in fact, the most commonly encountered leech on turtles in the northern United States and Canada, the known turtle hosts including *Chelydra serpentina*, *Chrysemys picta*, both the subspecies *belli* and *marginata*, *Sternothaerus odoratus*, *Graptemys geographica*, *Pseudemys scripta*, *Clemmys guttata*, and *Emydoidea blandingi*.

Detailed behavioral and cytological studies on the process of spermatophore implantation were made by Myers (1935) and Whitman (1891). *Placobdella parasitica* mates so readily that sometimes within minutes after collection one or more spermatophores can be found on their backs. On one occasion each I observed in the laboratory interspecific matings of this species with *Batracobdella picta* and *Haemopis marmorata*.

Adult *P. parasitica* leave their turtle hosts to become free-living during the breeding season in late summer, at which time and only then they are among the most commonly collected leeches in shallow water. At other times of the year they are usually on the hosts. Free-living brooding individuals have been found by myself and others on the following dates: 15 March, 3 May, 1 June, 22, 23, and 31 July, 1, 6, 10, 13, and 26 August, 30 November, and 19 December, which suggests that although they can breed from early spring until early winter, they breed predominantly in July and August. The thin-walled cocoons are deposited on the substrate, never attached to the venter of the parent. The cocoons are then covered tenaciously and ventilated by the parent, which does not leave them until the eggs hatch, except on rare occasions. After the eggs hatch, the young attach themselves by their hind suckers to the venter of the adult and are then carried about freely. Usually the parent with attached young remains free-living, but brooding individuals rarely occur on turtles. It is, however, improbable that the cocoons are ever attached to the host itself.

The species is able to lose 92 percent of the water in the body and still survive (Hall, 1922).

Distribution (Fig. 22). *Placobdella parasitica* is abundantly and widely distributed throughout north-central and eastern United States and southern Canada, being especially abundant in the Great Lakes region. This species, which is not yet known from the western states, is uncommon in the southern United States and most of Canada. In Canada its range is probably limited by the availability of its turtle hosts, especially the snapping turtle (*Chelydra serpentina*). The somewhat dubious record from Nicaragua (Moore, 1898) remains unconfirmed.

Published records believed valid include Alberta (Moore, 1964), Saskatchewan (Oliver, 1958), Ontario (Moore, 1906, 1922, 1936; Faull, 1913; Ryerson, 1915), Colorado (Herrmann, 1970), South Dakota (Moore, 1898), Nebraska (Ward, 1902), Kansas (Castle, 1900a), Iowa (Mathers, 1948), Minnesota (Moore, 1912), Wisconsin (Cahn, 1915; Bere, 1931; Sapkarev, 1968), Michigan (Hankinson, 1908; Miller, 1937; Kenk, 1949; Sawyer, 1968), Illinois (Castle, 1900a; Moore, 1901), Indiana (Moore, 1898, 1920), Ohio (Miller, 1929), Pennsylvania (Moore, 1912), New York (Moore, 1923; Barrow, 1953), Connecticut (Verrill, 1874a; Barrow, 1953), Massachusetts (Whitman, 1891; Castle, 1900a), Maine (Verrill, 1874a), southeastern Missouri (Moore, 1898; Meyer, 1937a), Tennessee (Moore, 1898), Louisiana (Sawyer, 1967), and Georgia (Patrick *et al.*, 1966). The discovery of specimens of this species from Arkansas, Kentucky, and South Carolina in the present study represents the first records for those states.

Placobdella ornata (Verrill, 1872)
(not *Placobdella ornata* Oka, 1929)

Clepsine ornata Verrill, 1872b:130; Verrill, 1874a:680; Verrill, 1874b:623; Verrill, 1875a:962; Moore, 1952:3.
Clepsine ornata var. *stellata:* Verrill, 1874a:681; Verrill, 1875a:962.
Clepsine ornata var. *rugosa:* Verrill, 1874a:681; Verrill, 1875a:964.
?*Glossiphonia parasitica* var. *rugosa:* Castle, 1900a:51, figs. C, 33.
Placobdella rugosa: Moore, 1901:487, figs. 2-3; Ward, 1902:278; Moore, 1906:157; Hankinson, 1908:232; Moore, 1912:86, figs. 6, 9; Andrews, 1915:200; Ryerson, 1915:166; Hankinson, 1916:118; Moore, 1918:654; Moore, 1920:90; Kraatz, 1921:150; Moore, 1922:8; Moore, 1923:15; Mullin, 1926a:36, pl. IV, fig. 3; Bere, 1929:177; Miller, 1929:10, fig. 5; Rawson, 1930:35; Bere, 1931:437; Moore, 1936:113: Meyer, 1937a:249; Meyer, 1937b:118; Miller, 1937:85; Richardson, 1943:90; Mathers, 1948: 397, pl. 2; Pawlowski, 1948:318; Kenk, 1949:38; Moore, 1952:3; Pennak, 1953:315; Meyer and Moore, 1954:84; Mann, 1961b:159; Patrick *et al.*, 1966:343.
Haementeria (*Placobdella*) *rugosa:* Autrum, 1936:61.

Placobdella ornata: Moore, 1952:3, figs. 1-3; Moore, 1959:550; Mann, 1961b: 159; Moore, 1964:1; Thomas, 1966:202; Sawyer, 1967:33; Sapkarev, 1968: 226; Sawyer, 1968:228; Scudder and Mann, 1968:208; Herrmann, 1970:5.
?*Placobdella multilineata:* Beck, 1954:74.

Description (Fig. 3E). *Placobdella ornata* is best distinguished by a wide brown band down the middorsal line, interrupted four or five times between the neck and anus. Characteristically, the dorsum is so heavily papillated as to appear warty. Each annulus has about 16-20 papillae of various sizes, being larger and more conspicuous in the posterior third of the back and toward the middorsal line. However, in some specimens a metameric pattern of papillae, somewhat like that of *P. papillifera*, can be distinguished: a single middorsal row and two pairs of paramedial rows of larger white-tipped papillae located on every third annulus. The marginal pigment pattern found in most members of the genus is the only pigment pattern on the dorsum; it is a mixture of faint green, dark brown, cream, and sometimes yellow and rusty brown. In some specimens a concentration of brown pigment occurs in the head and neck regions, superficially resembling the accessory eyes of *P. hollensis*. Although the venter, and in some specimens the dorsum, is generally mottled by numerous irregularly spaced, fine dark chromatophores, it is common to find individuals which have few or no such chromatophores. The general shape of resting specimens is flattened and lanceolate, but some unusually large individuals may be ovate-lanceolate with convex backs. Most individuals lack small papillae on the hind sucker, but occasionally a warty specimen may have conspicuous papillae on the sucker, resembling those found on *P. papillifera*.

Ecology. Although other hosts have been reported, such as the rock bass (*Ambloplites rupestris*) by Moore (1906) and the baldpate (*Marcea americana*) and the coot (*Fulica americana*) by Moore (1966a), this leech feeds predominantly on turtles, but not to the extent of *P. parasitica*. The known turtle hosts include *Chelydra serpentina*, *Sternothaerus odoratus*, *Chrysemys picta*, *Trionyx spiniferus*, *Trionyx muticus*, *Pseudemys scripta*, and *Terrapene carolina*.

Much less is known about reproduction for this common species than for *P. parasitica*. Insemination is by hypodermic injection of spermatophores, and it is not uncommon to find recently collected individuals with spermatophores attached dorsally and sometimes ventrally. It appears to breed somewhat earlier than *P. parasitica*. Brooding individuals have been found by myself and others on the following dates: 22 April, 21 May, 6 (three times) and 28 June, 1 (twice), 6, 7, 13, 25, and 30 July, 7, 14, 15, and 18 August, 1 September, and 5 October, which suggests that although they can breed from spring until autumn, they breed predominantly during June, July, and August. Unlike *P. parasitica*, *P. ornata* is commonly encountered off the turtle hosts at any time of the year.

On 22 April, 13 days after it was collected from a small stream in southeastern Michigan (water 12°C), a large (4 cm) individual deposited 80 eggs which were tightly enveloped in five sheath-like mucoid capsules or membranes, containing 26, 25, 19, 7, and 3 eggs and attached to the glass substrate at only one point. Other authors have reported as many as 95 young for this species (Moore, 1964). Like *P. parasitica*, the parent covered the young until they hatched on 8 May, after which they were carried about on the underside.

Distribution (Fig. 23). *Placobdella ornata* is abundantly and widely distributed throughout the northern United States and Canada. It is represented in the southern states by the common, closely allied form, *P. multilineata*, but its occurrence in the far western states has not been well documented. Published records believed valid include Alberta (Bere, 1929; Moore, 1964), British Columbia (Scudder and Mann, 1968), Manitoba (Meyer and Moore, 1954), Ontario (Moore, 1906, 1936; Ryerson, 1915; Rawson, 1930; Meyer, 1937b; Thomas, 1966), Quebec (Moore, 1922), Nova Scotia (Moore, 1922), Colorado (Verrill, 1874b; Herrmann, 1970), Kansas (Castle, 1900a), Nebraska (Ward, 1902), Iowa (Mathers, 1948), Minnesota (Moore, 1912), Wisconsin (Andrews, 1915; Bere, 1931; Sapkarev, 1968), Michigan (Hankinson, 1908, 1916; Miller, 1937; Kenk, 1949; Sawyer, 1968), Missouri (Meyer, 1937a), Illinois (Castle, 1900a; Moore, 1901; Baker, 1922), Indiana (Moore, 1920), Ohio (Moore, 1906; Kraatz, 1921; Miller, 1929), Pennsylvania (Moore, 1906), New York (Moore, 1923; Barrow, 1953), Connecticut (Verrill, 1874a; Barrow, 1953), and Massachusetts (Castle, 1900a). Reports of forms closely resembling and possibly identical with this species have also been reported from Utah (Beck, 1954), New Mexico (Verrill, 1875a), and Mexico (Caballero, 1940).

Some of the published records for this species may have been confused with *P. multilineata*, *P. papillifera*, or *P. hollensis*, all of which closely resemble *P. ornata*.

Placobdella papillifera (Verrill, 1872)

Clepsine papillifera Verrill, 1872b:130; Verrill, 1874a:683; Verrill, 1875a: 965; Moore, 1952:3.
Placobdella papillifera: Moore, 1952:3; Meyer and Moore, 1954:81; Mann, 1961b:159; Moore, 1964:1; Moore, 1966a:10.

Description (Fig. 3F). *Placobdella papillifera*, which varies from deep green to light brown, superficially resembles the more common *P. ornata*, from which it differs in several important respects. The body shape of the latter is generally flattened and lanceolate, whereas the body of *P. papillifera* is ovate-lanceolate and more convex dorsally. The papillae of *P. papillifera* are large, white-tipped, pointed, and restricted to five (sometimes seven) metameric rows. Any other papillae on the dorsum are irregularly arranged, small, and inconspicuous. In contrast, the papillae of *P. ornata* are moderately large, rounded, and occur over most of the back in no obvious metameric pattern, each annulus having 16-20 papillae. In some *P. ornata*, however, some papillae are larger, more white-tipped, and arranged in five metameric rows, thus resembling *P. papillifera*. In such specimens the most reliable characters for identification are the dorsal and ventral pigmentation patterns. Preserved *P. papillifera* have a continuous, or sometimes slightly interrupted, dark middorsal stripe encompassing the median row of papillae. On either side of this stripe and medial to the next pair of papillae is a characteristic whitish or somewhat bluish stripe which fuses with its counterpart at the neck, demarcating the anterior end of the brownish medial stripe. This stripe is rarely undeveloped, so that a single bluish stripe covers most of the region between the first pair of paramedial papillae, including the medial row. There is another, usually less distinct, pair of bluish stripes just medial to the second pair of papillae. In the neck region there is usually a suggestion of a white band which is not found in *P. ornata*. The latter has a characteristic dark brown interrupted middorsal stripe, but it is usually not flanked on either side by a white or bluish stripe and usually has no other indications of longitudinal pigmentation. Ventrally, the pigmentation of *P. papillifera* is characterized by two pairs of wide bluish longitudinal stripes, somewhat resembling the ventral stripes found in *P. parasitica*. There are no small dark chromatophores so characteristic of *P. ornata*, nor is there a midventral stripe. On the hind sucker of *P. papillifera* is a single row of papillae of uniform size but subject to variation between individuals.

Ecology. Individuals with young attached by their hind suckers to the ventral surface of the parent were found in southern Michigan on 28 May and 6 June 1967 (water 21°C). The only known host is the musk turtle (*Sternothaerus odoratus*).

Distribution (Fig. 24). Published records for *P. papillifera*, a poorly known species that is well represented in the present study from southern Michigan, had previously included only Alberta (Moore, 1964), Manitoba (Meyer and Moore, 1954), and Connecticut (Verrill,

1872b). It is much more common than published accounts would indicate and has probably been confused often with the remarkably similar *P. ornata*. The present records for Illinois and Michigan are the first for those states and, indeed, are the first well-established records for the United States.

Placobdella hollensis (Whitman, 1892)

Clepsine hollensis Whitman, 1892:385, pls. 39-40; Graf, 1899:224, fig. 110; Castle, 1900a:53.
Placobdella hollensis: Moore, 1906:157; Moore, 1912:94, fig. 11; Moore, 1918: 654; Mullin, 1926a:36, pl. IV, fig. 1; Mathers, 1948:397; Pennak, 1953: 315, fig. 200E; Moore, 1959:550; Mann, 1961b:159; Sawyer, 1968:228.
Haementeria (Parabdella) hollensis: Autrum, 1936:81.
Parabdella hollensis: Meyer and Moore, 1954:66.

Description (Fig. 3D). *Placobdella hollensis* has two pairs of variable dark concentrations of pigment situated metamerically behind the single functional pair of eyes, giving the false impression of three pairs of eyes (Fig. 17C). The checkered but somewhat variable dorsal pigment pattern consists essentially of a thick reddish-brown middorsal band, interrupted five or six times between the neck and the anus by broader squarish cream-colored patches about four annuli long. On either side of this middorsal band is a narrower band which is contiguous with the paired pigment concentrations of the false eyes. These bands are interrupted by squarish cream-colored patches in such a way that the brown portion flanks the white patches of the middorsal band. This complementation of pigment gives the center of the back a checkered appearance. The region between these bands and the margins is patternless, being irregularly pigmented with reddish brown.

Placobdella hollensis is similar to *P. ornata*, but unlike most *P. parasitica* and *P. papillifera*, in having numerous small dark chromatophores scattered more or less irregularly on the ventral surface. The chromatophores of *P. hollensis* differ slightly from those of *P. ornata*. The former contains a reddish-brown pigment, and the latter contains a black or dark brown pigment.

Placobdella hollensis is sometimes difficult to distinguish from *P. ornata*, but the former has few, if any, papillae along the center of the back, has a more or less checkered dorsum with light reddish-brown or brick-red pigment, is rarely if ever greenish, has metamerically arranged pigment concentrations or false eyes separated by two complete annuli, has a distinct flattened ribbon-shaped body, and swims readily when adult. It was common to find numerous individuals attached to the bottom of a canoe after a short trip around small leaf-bottomed ponds. *Placobdella ornata,* on the other hand, is characterized by having numerous dorsal papillae, especially prominent along the middorsal line, is usually dark brown mixed with green, does not have a checkered pattern on the dorsum, has a somewhat thicker and wider body, and rarely swims and then only as a juvenile. If pigment concentrations are present in the head region of *P. ornata,* they are not arranged metamerically, being separated at the most by one complete annulus. *Placobdella ornata* was commonly encountered on turtles, but *P. hollensis* was always free-living. The only known host, the painted turtle (*Chrysemys picta*), was reported in the original description.

Ecology. Almost nothing has been reported on reproduction in this species. On 22 May 1967 in southeastern Michigan (water 18°C) a large (3.5 cm) individual was found covering 193 eggs on a floating log, which did not have the leech the day before. The three membranous cocoons contained 77, 62, and 54 bright yellow eggs in early blastula stage. Judged from the eggs still in the ovisacs, at least one or more cocoon would have been deposited, making a total of well over 200 eggs. This would suggest that *P. hollensis* is perhaps our most fecund *Placobdella,* comparable to *Theromyzon.* On the same day this individual and seven other *P. hollensis,* ranging in size from 0.9 to 2.8 cm long, were placed together in a collecting vial. Within an hour the 3.5-cm specimen had five spermatophores implanted on its back, and a 2.4-cm specimen had one, but the six others had no implanted spermatophores at all. The possibility that the largest individuals of *P. hollensis* are most likely to be inseminated with spermatophores, a phenomenon observed also in *P. ornata* and *P. parasitica,* needs to be investigated. On other occasions spermatophores were also found on the ventral as well as the dorsal surfaces, once even in the genital region.

Distribution (Fig. 25). Published records believed valid for *P. hollensis,* a species well represented in the present study from southern Michigan, include only Minnesota (Moore, 1912), Iowa (Mathers, 1948), Michigan (Sawyer, 1968), and Ontario (Moore, 1906), but it is probably much more common than published accounts would indicate. Some of the published accounts for *P. ornata* and *P. papillifera* may have been confused with this species.

Placobdella montifera Moore, 1906
(not *Placobdella carinata* Diesing, 1858)

?*Glossiphonia trisulcata:* Baird, 1869:317 (the specimen I examined in the British Museum resembles *P. montifera*).

Clepsine papillifera carinata Verrill, 1874a:683; Moore, 1952:4.
Hemiclepsis carinata: Moore, 1901:498, fig. 5.
Placobdella montifera Moore, 1906:156; Moore, 1912:88, figs. 5, 10; Ryerson, 1915:166; Moore, 1918:652; Moore, 1920:89; Moore, 1924b:23; Mullin, 1926a:36; Miller, 1929:10, fig. 3; Bere, 1931:437; Moore, 1936:113; Meyer, 1937a:249; Meyer, 1937b:118; Miller, 1937:85; Eddy and Hodson, 1945:29; Meyer, 1946a:237; Mathers, 1948:397, pl. 2; Moore, 1952:4, figs. 6-7; Pennak, 1953:315, fig. 200B; Oliver, 1958:163; Moore, 1959:549; Harms, 1960:698; Paloumpis and Starrett, 1960:416; Mann, 1961b:158; Moore, 1964:13; Patrick *et al.*, 1966:343; Thomas, 1966:202; Carlson, 1968:164; Meyer, 1968:11; Sapkarev, 1968:226.
Haementeria (Placobdella) montifera: Autrum, 1936:64.

Description (Fig. 3A). *Placobdella montifera* has three keel-like ridges on the dorsum, one middorsal ridge, a pair of paramedials (corresponding to the outer rather than the inner pair of paramedial papillae in *P. papillifera*), and a characteristic narrow constriction in the neck region, setting off the wide head from the rest of the body. The body is almost rectangular, being somewhat more narrow anteriorly. Each ridge in *P. montifera* is composed of uniformly large, pointed tubercles which, unlike *P. papillifera*, which it vaguely resembles, occur on every annulus. The area between the ridges is relatively smooth, having only a few very small inconspicuous papillae, in contrast to *P. ornata*, in which each annulus has numerous papillae from one margin to the other and no comparable smooth areas.

In *P. montifera* an inconspicuous longitudinal row of papillae is sometimes found between the middorsal and the paramedial ridge and near the margins on either side respectively, but (except for three or four large anal tubercles corresponding to the first pair of paramedial papillae in *P. papillifera*) these papillae do not reach the size found in *P. papillifera* or *P. ornata*.

Ecology. Various authors have reported that *P. montifera* will attack aquatic worms, insect larvae, mussels, frogs, toads, and fish, but the only specific host records have been fish. It has been found twice on pumpkinseed (*Lepomis gibbosus*) and once each on bass (*Micropterus salmoides, M. dolomieui*), gar (*Lepisosteus osseus*), black bullhead (*Ictalurus melas*), silver redhorse (*Moxostoma anisurum*), and carp (*Cyprinus carpio*) (Hoffman, 1967).

Very little is known about reproduction in this uncommon species except that insemination is by hypodermic injection of spermatophores and the young are carried about by the parent (Moore, 1912; Mathers, 1948).

Distribution (Fig. 25). Published records believed valid include British Columbia (Clemens *et al.*, 1939), Saskatchewan (Oliver, 1958),

Ontario (Moore, 1906, 1924, 1936; Ryerson, 1915; Thomas, 1966), Minnesota (Moore, 1912), Wisconsin (Bere, 1931; Sapkarev, 1968), Michigan (Verrill, 1874a; Miller, 1937), Iowa (Mathers, 1948; Carlson, 1968), Kansas (Harms, 1960), Missouri (Meyer, 1937a), Illinois (Moore, 1901; Paloumpis and Starrett, 1960), Indiana (Moore, 1920), Ohio (Miller, 1929), and Georgia (Patrick *et al.*, 1966).

Placobdella pediculata Hemingway, 1908

Placobdella pediculata Hemingway, 1908:527, figs. 1-3; Hemingway, 1912:33, pls. C-E; Moore, 1912:90, figs. 13-18; Ryerson, 1915:169; Moore, 1918: 653; Mullin, 1926a:36, pl. IV, fig. 4; Bere, 1931:437; Meyer, 1937a:249; Mathers, 1948:397; Richardson, 1949:85; Pennak, 1953:315, figs. 200C, D; Moore, 1959:560; Branson and Amos, 1961:53; Mann, 1961b:158.
Haementeria (*Placobdella*) *pediculata*: Autrum, 1936:79, fig. 52.

Description (Fig. 3B). *Placobdella pediculata*, the only truly non-papillated *Placobdella* in North America, has a conspicuously stalked caudal sucker, although individuals shorter than 1 cm lack the long peduncle. Hemingway (1908, 1912) showed that the peduncle arises during development, its size corresponding to its depth into the tissue of the host. Both young and adult *P. pediculata* differ from *Actinobdella*, the only other North American glossiphoniid that has a conspicuously stalked caudal sucker, in lacking dorsal papillae and in having the anus at a uniquely anterior position, XXIII/XXIV, rather than at the usual position, XXVII/XXVIII.

Ecology. This leech usually imbeds its sucker in or near the isthmus below the gill chamber of the drum (*Aplodinotus grunniens*). Other authors have found this species attached posteriorly in the region of the dorsal fins (Mullin, 1926). In spite of unconfirmed reports of its temporary attachment to turtles (Mathers, 1948) and to fish of the families Cyprinidae and Catostomidae (Branson and Amos, 1961), there is every reason to believe that *P. pediculata* has a high degree of host specificity for the drum. Apart from the report of finding adults on the host in August and young (1 cm) in September (Hemingway, 1912), and one report of finding it free-living under a rock (Bere, 1931), little is known about the habits of this interesting leech.

Distribution (Fig. 24). This species is known primarily from the midwestern states west and southwest of the Great Lakes: Minnesota (Lake Pepin, Hemingway, 1912), Wisconsin (northeastern lakes, Bere, 1931), Illinois (Henry and Peoria, Moore, 1912), Iowa (Okoboji region, Mathers, 1948), Missouri (Cape Girardeau County, Meyer, 1937a), and Oklahoma (Lake Texoma, Branson and Amos, 1961). It was once reported from Maine (DeRoth, 1953).

Placobdella multilineata Moore, 1953

Placobdella multilineata Moore, 1953:1, pl. 1, fig. 1; Moore, 1959:550; Mann, 1961b:159; Sawyer, 1967:33; Meyer, 1968:11.
Placobdella rugosa southern variety: ?Pearse, 1936:181; Moore, 1953:4.

Remarks. In the southern states *P. ornata* is replaced by the closely related *P. multilineata*, which, like *Philobdella gracilis*, may occur as far north as southern Illinois. *Placobdella multilineata* has a continuous brown stripe, whereas *P. ornata* has an interrupted brown middorsal stripe. However, the finding of a few individuals of *P. multilineata* with interrupted stripes in Louisiana (Sawyer, 1967) and the discovery in the present study of an occasional specimen of *P. ornata* with a continuous middorsal stripe in Michigan and Illinois suggest that the continuity of this stripe alone does not necessarily distinguish the two forms. Judged from some Louisiana specimens, *P. multilineata* differs from *P. ornata* in having much fewer and smaller papillae arranged in five longitudinal series, thus resembling *P. papillifera*. In addition, both the dorsal and ventral surfaces of *P. multilineata* have suggestions of longitudinal pigment patterns, notably fine dark chromatophores arranged between the longitudinal muscle bands. Nonetheless, the differences between these two species remain so slight that *P. multilineata* may eventually prove to be a southern subspecies of *P. ornata*, a problem which needs investigation.

GENUS *HELOBDELLA* R. BLANCHARD, 1896

The genus *Helobdella*, which has its center of distribution in South America, was recognized as a natural group, separate from the genus *Glossiphonia*, by R. Blanchard in 1896. Over 21 recognized species and subspecies are known from South America (Weber, 1913, 1915; Pinto, 1923; Autrum, 1936; Cordero, 1937; Ringuelet, 1943-45), but only four or five recognized species are known from North America (Moore, 1906, 1959). Some of the North American species may be represented in South America under the same or different names. On the whole, *Helobdella* is the most taxonomically confusing group in the Americas, primarily because of the unsettled problem of polymorphism in the *triserialis* complex of species in South America and in the *fusca* group, its North American counterpart.

Ringuelet, who critically examined the South American *Helobdella* (1943, 1944a, 1944b, 1945), recognized 17 species in addition to five subspecies of *triserialis*, at least one of which, *H. triserialis lineata*, clearly belongs to the *fusca* group. The latter group was examined by Moore (1906), who distinguished three varieties, *fusca*, *lineata*, and *papillata*. In 1906 Moore felt that these forms were connected by a continuous variation of characters, but by 1959 he recognized each of these as a species, in addition to adding a fourth related species, *H. punctatolineata*, from Puerto Rico (1939).

In the present study the three species *H. fusca*, *H. papillata*, and *H. lineata* were fairly often encountered, proving to be consistently recognizable forms. Three questionable forms of *H. fusca* were also encountered on only one or two occasions and will be discussed below.

Helobdella stagnalis (Linnaeus, 1758)

[A full European synonymy can be found in Harding (1910) and Autrum (1936).]

Hirudo stagnalis Linnaeus, 1758:649.
Clepsine modesta: Verrill, 1872b:129, fig. 2; Verrill, 1873b:388; Verrill, 1874a: 679; Verrill, 1875a:961; Moore, 1952:3.
Clepsine submodesta: Nicholson, 1873:493.
Clepsine minima: Sager, 1878:74.
Glossiphonia stagnalis: Moore, 1898:549; Castle, 1900a:21, figs. A, 4, 7-10, 12, 34; Moore, 1901:497; Ward, 1902:277; Moore, 1906:156, fig. 2; Moore, 1912:77, fig. 1; Ryerson, 1915:165; Moore, 1918:651; Moore, 1920:89; Kraatz, 1921:150; Moore, 1922:7; Mullin, 1926a:35; Miller, 1929:10, fig. 4; Miller, 1937:85; Richardson, 1942:68.
Clepsine bioculata: Graf, 1899:224.
Glossiphonia (Helobdella) stagnalis: Moore, 1922:9.
Helobdella stagnalis: Moore, 1923:15, pl. 1A; Moore, 1924b:22; Richardson, 1925a:361; Richardson, 1928:406; Bere, 1929:177; Rawson, 1930:35; Bere, 1931:437; Moore, 1936:113; Meyer, 1937a:249; Moore, 1937:118; Townes, 1937:167; Richardson, 1943:89; Mathers, 1948:397, pls. 1, 4; Pawlowski, 1948:331; Kenk, 1949:38; Moore and Meyer, 1951:59; Moore, 1952:3; Pennak, 1953:314, fig. 200A; Beck, 1954:74; Meyer and Moore, 1954:68; Oliver, 1958:163; Moore, 1959:548; Fredeen and Shemanchuk, 1960:733; Paloumpis and Starrett, 1960:416; Mann, 1961b:156; Hilsenhoff, 1963:252; Moore, 1964:1; Moore, 1966a:10; Patrick *et al.*, 1966:343; Thomas, 1966: 202; Sawyer, 1967:35; Carlson, 1968:164; Sapkarev, 1968:226; Sawyer, 1968:228; Clifford, 1969:583; Gates and Moore, 1970:45; Herrmann, 1970: 5; Mason *et al.*, 1970:R323.
?*Erpobdella stagnalis:* Oliver, 1958:164.

Description (Fig. 5C). *Helobdella stagnalis* is the only species in North America with a brown horny scute in the neck region (Fig. 17E). On the rare occasions when the scute is missing, such as after poor preservation, the species could be confused with *H. elongata*, a small whitish allied form with which it is often associated. *Helobdella stagnalis* is usually opaque, larger, thicker, and relatively wider than *H. elongata*, the body of which is translucent, narrow, and excessively flattened, almost ribbon-like. Cleared specimens of these species are easily distinguished because *H. stagnalis* has six pairs of crop caeca, whereas *H. elongata* has only one pair.

Variation. Usually *H. stagnalis* is about 10-12 mm long when mature, but the size may vary considerably, occasionally up to 20 mm. Individuals which have been preserved quickly without prior narcotization are especially wide, the width being about three-quarters of the length. The amount of pigment may vary considerably, both between populations and within a population. Usually individuals are almost completely unpigmented, but some may contain so many diffusely arranged black chromatophores, both dorsally and ventrally, that the animal appears gray or even blackish. In some individuals the scute is large and almost triangular, being broadest anteriorly, whereas in others it is small and rod-shaped. In still others it is small and disc-shaped, sometimes so small that it may go unnoticed.

Ecology. The species feeds on small oligochaetes, aquatic insects, and possibly other leeches rather than on snails (Hilsenhoff, 1963; Thut, 1969). In the laboratory I was unable to get it to feed on the snails *Physa, Stagnicola,* or *Helisoma.* In early spring *H. stagnalis* was commonly found attached to other leeches, *Batracobdella picta, Haemopis grandis, H. marmorata,* and *Macrobdella decora,* but, unlike *Glossiphonia complanata,* it was never actually seen feeding on them.

Little is known about reproduction in this species. The process of egg-laying in particular has not been described previously. The following observations of egg-laying were made in early May on an individual captured four days previously under a rock in a small stream in Washtenaw County, Michigan. When first observed, the leech was attached to the side of a glass jar, through which its venter could be easily seen. Forty flesh-colored eggs were attached, as if glued, to the posterior third of its venter. From the eggs that could still be seen inside the leech it was apparent that it was in the process of laying its eggs, the process probably beginning around dusk. The leech when first discovered was apparently in an interim period between egg-laying, a period characterized by posteriorly moving undulations or ventilatory movements (Fig. 7A) that continued for less than a minute before the following egg-laying repertoire was resumed.

The thin lateral margins of the posterior third of the leech inflected downward and inward toward the ventral center, creating a trough around the eggs that had already been laid (Fig. 7B), followed immediately by the arching of the back so as to create a cavity around these eggs (Fig. 7C). During the brief period for which the animal held this position, the muscles of the body seemed to be forcing the anterior pair of eggs, which were moving freely about in the ovisac, into the oviduct, a chamber immediately posterior and internal to the gonopore. Having the eggs situated in the oviduct may be a prerequisite for their expulsion to the outside, the eggs in the oviduct at this stage becoming the contents of one protective sac or cocoon. Close examination of the translucent body of the leech, which by now was in another interim period displaying only ventilatory movements, revealed that within the ovisacs the most posterior eggs, which had been forced anteriorly by the arched back and straining motions, moved passively back to their original positions. The two eggs next to be laid were clearly seen in the oviduct, the other eggs lying posterior to them, deep in the ovisacs.

Actual egg-laying proceeded as follows: the head went through a searching movement accompanied by strong ventral flexure (Fig. 7D). As this flexure increased and the body rolled into a ball (Fig. 7E), the head moved toward the hind sucker, then quickly twisted to one side, while the gonopore became placed just cephalad to the previously laid eggs. For about ten seconds the head went into a vigorous circular movement, again reminiscent of muscular straining (Fig. 7F), accompanied by several thrusting movements of the gonopore region, apparently directed toward egg extrusion. Next the body, which had been only slightly arched during the actual egg extrusion, arched fully so that the head was directly under it (Fig. 7G). This position was held for about five seconds and was accompanied by vigorous straining movements. It was difficult to see what the head was doing, but it may have been "shaping" the cocoon, as observed in *Erpobdella punctata* by Sawyer (1970a). Then began slight oscillations (Fig. 7H), which soon increased until the arched body was unrolled into the initial extended position in which ventilation occurs (Fig. 7A). In having the oral sucker attached to the substrate during ventilation, *H. stagnalis* resembles *Theromyzon meyeri* but differs from *Erpobdella punctata*.

Close examination showed that the oviduct which had held the two eggs was now empty, and a count revealed that the exposed eggs had increased by two. The entire ritual described above was repeated successively with very little deviation except for possible variation in the number of eggs laid. The leech laid a total of 60 eggs in one night, the more posterior sacs containing the most eggs (Fig. 6E); the eggs hatched in about five days.

In a later study in a small permanent pond in southeastern Michigan, each individual was found to have laid an average of 8.4 egg sacs or cocoons (Fig. 6C), the larger individuals having laid more

(Fig. 6F). Each cocoon contained an average of 4.2 eggs (Fig. 6D). No correlation was found between increased size of adult and increased number of eggs per cocoon; the increase in the number of eggs produced by larger individuals results only from the production of additional cocoons.

Brooding individuals were encountered in the present study from early April to August in water over 21°C, suggesting that in Michigan, as in England (Mann, 1957a), more than one generation may be produced each year.

Distribution (Fig. 26). In addition to being found on every continent except Australia, *H. stagnalis* occurs abundantly throughout the northern United States and Canada but is less common in the southern United States. Published records believed valid include Northwest Territories (Moore and Meyer, 1951; Meyer and Moore, 1954), southern tip of the Georgian Bay, British Columbia (Clemens *et al.*, 1939; Meyer and Moore, 1954), Alberta (Bere, 1929; Moore, 1964; Clifford, 1969), Saskatchewan (Oliver, 1958), Ontario (Moore, 1906; Faull, 1913; Moore, 1922, 1924b; Rawson, 1930; Moore, 1936; Meyer, 1937b; Meyer and Moore, 1954; Thomas, 1966), Quebec (Moore, 1922), Newfoundland, St. Pierre (Pawlowski, 1948), Nova Scotia (Pawlowski, 1948; Meyer and Moore, 1954; Gates and Moore, 1970), New Brunswick, Prince Edward Island (Richardson, 1943), Washington (Thut, 1969), Oregon (Mason *et al.*, 1970), California (Verrill, 1875a; Gee, 1913), Arizona (Verrill, 1874b), Utah (Verrill, 1875a; Beck, 1954), Colorado (Verrill, 1874b; Herrmann, 1970), Nebraska (Verrill, 1874b), Iowa (Mathers, 1948; Carlson, 1968), Minnesota (Moore, 1912), Wisconsin (Bere, 1931; Hilsenhoff, 1963; Sapkarev, 1968), Michigan (Hankinson, 1908; Miller, 1937; Kenk, 1949; Sawyer, 1968), Illinois (Richardson, 1925a, 1928; Paloumpis and Starrett, 1960), Indiana (Moore, 1920), Ohio (Moore, 1906; Kraatz, 1921; Miller, 1929), Pennsylvania (Moore, 1906, 1912), New York (Moore, 1922, 1923; Barrow, 1953), New Jersey (Castle, 1900a), Connecticut (Verrill, 1874a; Barrow, 1953), Massachusetts (Moore, 1898; Castle, 1900a; Weston and Turner, 1917), southeastern Missouri (Meyer, 1937a), Georgia (Patrick *et al.*, 1966), and Florida (Verrill, 1874a). The finding of this species in South Carolina in the present study is the first record for that state.

Helobdella lineata (Verrill, 1874)
(not *Hirudo lineata* O. F. Müller, 1774)

Clepsine papillifera var. *lineata* Verrill, 1874a:683; Ward, 1902:277; Moore, 1952:3.

Glossiphonia lineata: Moore, 1898:549; Moore, 1901:493; Ward, 1902:277; Moore, 1952:10, fig. 5.
Glossiphonia fusca lineata: Moore, 1906:159; Baker, 1924:109; Moore, 1952:10.
Glossiphonia fusca: Moore, 1912:80, fig. 3; Ryerson, 1915:165; Moore, 1920:89; ?Kraatz, 1921:150; (?part) Moore, 1922:7; Mullin, 1926a:48; Miller, 1929:10; Meyer, 1937a:249.
Helobdella fusca: (part) Moore, 1918:652; Mathers, 1948:397, pl. 1; ?Herrmann, 1970:5.
Glossiphonia (Helobdella) fusca: (?part) Moore, 1922:9.
Helobdella triserialis lineata: Ringuelet, 1943:229, fig. 3; Sawyer, 1967:34.
Helobdella lineata: Moore, 1952:3; Moore, 1959:549; Paloumpis and Starrett, 1960:416; Mann, 1961b:156; Sapkarev, 1968:226; Sawyer, 1968:228; Herrmann, 1970:2.

Description (Figs. 4B, C). *Helobdella lineata* has three series of black-tipped papillae and four series of metameric white dots, which may disappear in preserved specimens. The two paramedial series of papillae are shorter than the middorsal row, which can extend anteriorly at least to the clitellar region. The papillae are positioned on every third annulus of complete segments and have a somewhat staggered, rather than linear, arrangement. These papillated annuli also bear white metameric dots, of which there is a row on either side of the middorsal row of papillae and a row just external to each row of paramedial papillae. Two rows of dots nearest to the middorsal line extend farthest, almost to the neck region and well beyond the middorsal papillae. The species is about 8-10 mm long but may be as long as 20 mm, especially in the southern part of the range.

Variation. On the rare occasions when the dorsal papillae are absent, the species can usually be distinguished from *H. fusca* by the four series of metameric dots instead of continuous longitudinal stripes. Some individuals of *H. lineata* had longitudinal whitish stripes among which vestiges of the four rows of metameric dots were discernible. In a few the dorsal papillae were missing or so inconspicuous that the leech resembled *H. fusca* to such a remarkable degree that externally there was no reliable way of separating the two species. In fact, it is the occurrence of these apparent intergrades which supports the possibility discussed below that *H. lineata* and *H. fusca* belong to the same polytypic species. If such is the case, the most widely distributed form would undoubtedly be the papillated *H. lineata*.

Remarks. This species was first recognized by Verrill (1874a) as *Clepsine papillifera* var. *lineata,* but Moore (1906), who mistakenly thought that the name *lineata* was preoccupied by *Hirudo lineata* O. F. Müller, 1774, gave it the name *Glossiphonia fusca* (Castle, 1900) *lineata.* Later (1952) he recognized it as a distinct species, *H. lineata*

(Verrill, 1874), separate from *H. fusca* (Castle, 1900), but he left open the possibility that the latter might be a polymorphic species. Moore was also aware of the closeness of *H. lineata* and *H. triserialis* but did not give them the same name because the male pore of the latter, according to R. Blanchard (1896a), was supposedly at XI/XII instead of XIIa1/a2, as found in *H. lineata*. Ringuelet (1943) in his excellent review of the polymorphic South American species *H. triserialis* (E. Blanchard, 1849) regarded *H. lineata* and *H. fusca* as varieties of one extremely variable, widely distributed species. After examining large numbers of South American *Helobdella*, Ringuelet concluded that Blanchard had been wrong in his original observation and that this error had been repeated by subsequent authors. He found in *H. triserialis*, as Moore did in *H. lineata*, that the gonopores are separated by one rather than two annuli.

Ringuelet was probably right in recognizing one widely distributed, variable species, *H. triserialis*, as encompassing *H. triserialis lineata*, but until the problem of polymorphism is settled in the *H. fusca* complex of species, the names for the American *Helobdella* used in this study will basically follow Moore (1959).

Distribution (Fig. 27). *Helobdella lineata*, which extends as far north as the lower Great Lakes, especially Lake Erie, is a warm-water species which becomes a dominant species of the lower Mississippi Valley. As the common species *H. triserialis*, it probably continues into Central and South America (Ringuelet, 1943). Published records believed valid, often as *H. fusca*, with which it has probably been confused many times, include Minnesota (Moore, 1912), Michigan (Sawyer, 1968), Iowa (Mathers, 1948), Nebraska (Verrill, 1874a; Moore, 1952), Illinois (Moore, 1901), Wisconsin (Baker, 1924; Sapkarev, 1968), Indiana (Moore, 1920), Ohio (Moore, 1906; Miller, 1929), Pennsylvania (Moore, 1906), Ontario (Moore, 1906), southeastern Missouri (Meyer, 1937a), Louisiana (Sawyer, 1967), and Mexico (Moore, 1898). It has not previously been reported from South Carolina, where it is common.

Helobdella fusca (Castle, 1900)

Glossiphonia fusca Castle, 1900a:34, figs. 13-18; Moore, 1918:652, fig. 997; (?part) Moore, 1922:7; Moore, 1936:113; ?Meyer, 1937b:118; ?Miller, 1937:90.

Glossiphonia fusca fusca: Moore, 1906:158, fig. 5.

Glossiphonia (Helobdella) fusca: (?part) Moore, 1922:9.

Helobdella fusca: ?Moore, 1924b:22; ?Bere, 1931:439; ?Mozley, 1932:244; Autrum, 1936:29, fig. 18; ?Townes, 1937:167; ?Kenk, 1949:38; Moore and Meyer, 1951:60; Pennak, 1953:314; Meyer and Moore, 1954:68; Moore, 1959:548, fig. 23.3; Mann, 1961b:156; Moore, 1966a:10; ?Patrick *et al.*, 1966:342.

Description (Figs. 4D-F). Unlike other *Helobdella*, *H. fusca* lacks dorsal papillae and has three major pairs of longitudinal whitish stripes, a paramedial, an intermediate, and a marginal pair. Usually the intermediate stripe is the longest, extending almost to the eye region. The marginals are the shortest and extend from the anus to the neck, separated from the margins of the body by a narrow longitudinal band of pigment. The paramedials usually fade out in the neck region, but in some cases they extend continuously to the eyes. Between the paramedials a pigmented middorsal band extends from the posterior third of the body up to the eyes. Immediately posterior to this band is a short but conspicuous white anal bar. Small metameric white dots resembling those found in the strongly papillated *H. lineata* can be seen, especially in the anal region.

Remarks. In addition to typical *H. fusca*, two other unpapillated color forms were encountered in Michigan: a whitish form with the dorsal surface completely lacking pigmentation (Fig. 4F), and a mottled form with a dorsal surface pattern consisting of irregularly spaced white blotches on a brownish background but without longitudinal stripes and middorsal band (Fig. 4E). In southeastern Michigan on 6 August 1966 brooding individuals of all three color forms were found in the same pond with no apparent intergradation of characters.

Ecology. Other workers have shown that this or a closely related species favors snails as food and could possibly help as a biological control agent for the snail-borne disease schistosomiasis (Chernin *et al.*, 1956; McAnnaly and Moore, 1966). In a central Michigan lake I dredged *H. fusca* from the bottom, still attached to the shells of one of its probable Michigan hosts, the snail *Helisoma*. Each brooding parent carries on its venter an average of six to seven capsules or cocoons, each containing about 15 eggs; the breeding season appears to be from June to August (Castle, 1900a; Moore, 1966).

Distribution (Fig. 27). *Helobdella fusca* appears to be a cold-water species with a more northern and eastern distribution than *H. lineata* and more likely to be found in lakes and larger ponds. The possibility that *H. fusca* replaces *H. lineata* in the Great Lakes region, where there is an apparent overlap of ranges, needs to be investigated. Because of the confusion over the names *lineata* and *fusca*, most reports in the literature are not reliable. The records include Northwest Territories (Moore and Meyer, 1951; Meyer and Moore, 1954), Alberta (Moore, 1964, 1966), Manitoba (Meyer and Moore, 1954), Ontario (Moore, 1906, 1936), Ohio (Moore, 1906), New Jersey (Castle, 1900a), and Massachusetts (Castle, 1900a). It has not previously been found in Michigan.

Helobdella elongata (Castle, 1900)

Clepsine nepheloidea: Graf, 1899:224 (inadequate description).
Glossiphonia elongata Castle, 1900a:39, figs. B, 23-27; Moore, 1906:158.
Glossiphonia nepheloidea: Moore, 1906:156; Moore, 1912:76; Ryerson, 1915: 165; Moore, 1918:651; Mullin, 1926a:35; Miller, 1929:10; Rawson, 1930: 35; Miller, 1937:90; Townes, 1937:167.
Helobdella nepheloidea: Moore, 1924b:22; Richardson, 1925a:348; Bere, 1931:439; Moore, 1936:113; Meyer, 1937a:249; Mathers, 1948:397, pl. 1; Pennak, 1953:314; Meyer and Moore, 1954:66; Hilsenhoff, 1964:139; Carlson, 1968:164.
Helobdella elongata: Autrum, 1936:28; Moore, 1959:548; Paloumpis and Starrett, 1960:416; Mann, 1961b:156; Patrick *et al.*, 1966:342; Sawyer, 1967:34; Sapkarev, 1968:226.
Glossiphonia nepheloidae: Miller, 1937:89.

Description (Fig. 5D). *Helobdella elongata* has a cylindrical worm-like body which is so unpigmented and translucent that the internal organs, especially the crop and the large gland cells, show through the body wall. Unlike other *Helobdella*, the margins of the body are nearly parallel, not much wider than the small hind sucker. The species has only one pair of crop caeca, whereas most *Helobdella* have six pairs. The South American species *H. michaelseni* R. Blanchard, 1900, which also has one pair of crop caeca, may represent the same species.

Remarks. Very little is known about its biology except that it feeds on aquatic insect larvae (Hilsenhoff, 1964). In Michigan I found the species carrying eggs from late May to early June (water 21°C).

Distribution (Fig. 28). *Helobdella elongata* is widely, but sporadically, distributed from the midwestern and Great Lakes states eastward and southward. Reliable published records include Ontario (Ryerson, 1915; Moore, 1924b; Rawson, 1930; Moore, 1936), Iowa (Mathers, 1948; Carlson, 1968), Minnesota (Moore, 1912), Wisconsin (Bere, 1931; Miller, 1937; Hilsenhoff, 1964; Sapkarev, 1968), Illinois (Richardson, 1925a; Paloumpis and Starrett, 1960), Ohio (Moore, 1906; Miller, 1929), Pennsylvania (Moore, 1906), New York (Barrow, 1953), Connecticut (Barrow, 1953), Massachusetts (Castle, 1900a; Moore, 1912), Missouri (Meyer, 1937a), Louisiana (Sawyer, 1967), and Georgia (Patrick *et al.*, 1966). This is the first record for Michigan.

Helobdella papillata (Moore, 1906)

Clepsine papillifera var. *b:* Verrill, 1874a:683; Moore, 1952:3.
Helobdella fusca var. *papillata* Moore, 1906:159; Moore, 1952:10, fig. 4.

Helobdella fusca: (part) Moore, 1918:652.
Helobdella papillata: Moore, 1952:3; Moore, 1959:549; Mann, 1961b:157; Meyer, 1968:8; Sawyer, 1968:228.

Description (Fig. 4A). *Helobdella papillata* has numerous rounded papillae that protrude conspicuously from its back, positioned on every third annulus in complete segments and arranged in five to seven longitudinal rows. In some specimens the papillae are black-tipped, and in others on either side there are two longitudinal series of white dots, just external to the medial and first pair of paramedial papillae respectively, resembling the dots in *H. lineata*. *Helobdella papillata* and *H. lineata* can be distinguished by the roughly papillated appearance of the former and the smaller, less numerous papillae of *H. lineata*. The rare occurrence of apparently intermediate forms suggests that *H. papillata* and *H. lineata*, as with *H. fusca*, represent one variable species.

Remarks. Originally described by Verrill (1874a) as *Clepsine papillifera* var. *b*, this species was later described independently as a variety of *H. fusca* (Castle, 1900) by Moore (1906), who gave it the name *H. fusca* var. *papillata*. Moore (1952) later examined Verrill's collection, discovered that the two species were the same, and established the name *H. papillata*. Very little is known about its biology, distribution, or, for that matter, phylogenetic position. In Michigan I found individuals with the young on 30 July and 8 August, suggesting that this species, like *H. fusca*, breeds in midsummer.

Distribution (Fig. 28). Published records believed reliable include Michigan (Sawyer, 1968), the Ontario and Ohio sides of Lake Erie (Moore, 1906), and possibly Connecticut (Verrill, 1874a). Its known range is considerably enlarged by the present study, which provides the first records for Minnesota and Illinois.

Helobdella transversa, new species

Type-Locality. Berrien County, Michigan (creek between North and Middle Lake Mere lakes). The holotype and paratype have been deposited in the Charleston Museum in South Carolina.

Description (Figs. 5A, B). On 28 May 1967 ten specimens of this undescribed species, only the diagnostic characters of which will be presented here, were found in two localities in Berrien County, Michigan, in association with *H. stagnalis* and *H. fusca*. The unpapillated dorsal surface is generally rusty brown, interrupted by metameric white bands which consist of eight to ten slightly raised white dots in various states of confluence. These white bands of dots tend to fade out at the anterior quarter of the body. The pattern is conspicuously transverse with no longitudinal pattern of white and rusty-brown stripes as in *H. fusca*, which it superficially resembles. The pigment fades quickly in ethanol, unlike that of *H. fusca*. They are all close to 10 mm long with a circular hind sucker 1 mm in diameter. The eyes (one pair) are well separated and positioned on the fourth annulus. The reconstruction of the digestive and reproductive systems of the types (Fig. 5B), based on a transverse series of a paratype at 10 μ and stained with eosin and haemotoxylin, is substituted for a lengthy verbal description of these systems, which are typical of the genus. Like most *Helobdella*, this species has six pairs of crop caeca. The male gonopore is positioned at XIa1/a2, and the obscure female opening is at XIIa2/a3, which is one annulus caudad.

GENUS *ACTINOBDELLA* MOORE, 1901

The genus *Actinobdella* was erroneously placed in the family Piscicolidae by Moore (1901), who was misled by the unusual six-annulate condition of the type-species, *A. inequiannulata*. The internal and external morphology of a second and third species described in 1906 and 1924, *A. annectens* and *A. triannulata* respectively, showed beyond doubt that the genus belonged to the family Glossiphoniidae. It most closely resembles *Placobdella* and *Batracobdella*, but the morphology and biology of this inadequately known and perhaps unnatural genus must be investigated before its true systematic standing can be determined. There is a remarkable resemblance between *A. annectens*, known only from the original description, and *Batracobdella phalera*, which has been reported on various occasions with numerous small papillae along the margins of the caudal sucker (Moore, 1906), with an unusually long posterior sucker (Ryerson, 1915), and was once found attached to the gill arch of a fish (Bere, 1931), all three characters reminiscent of *Actinobdella*.

Actinobdella have a large, almost hemispherical posterior sucker separated from the body by a narrow pedicel. Projecting into the sucker cavity a short distance from its inner margin is a circle of 30-60 retractile digitate processes with accessory adhesive gland ducts. The bodies of the individuals, which range in size from 1.5 to 12 mm, with an average about 9 mm, are only moderately wide, with more or less parallel sides, and are somewhat convex, never excessively flattened. Complete segments have three annuli, which in two species are further divided into six unequal annuli, an unusual condition for glossiphoniids. They also have a single pair of large eyes, either confluent or very close together, seven pairs of branching caeca and four pairs of intestinal caeca, diffuse salivary glands, a loosely folded epididymis, one to three series of dorsal papillae, and a mouth located far forward on the oral sucker.

This genus is represented in the present study by one poorly preserved individual labeled "H-51 Lake Chautauqua, Havana, Illinois, 23 April 1953" and dubiously identified "*Actinobdella inequiannulata.*" This is probably the same specimen reported by Paloumpis and Starrett (1960) from Lake Chautauqua. The elongated individual is widest at the anterior region, from which it tapers slightly to the posterior region (Fig. 1D). The body is mostly depressed rather than convex and has no dorsal papillae. It has about six unequal annuli per segment, but in the absence of external signs of metamerism it is difficult to be certain. There is one pair of confluent eyes. Nothing could be determined about its internal anatomy after clearing. The posterior sucker (Fig. 1E), which is on a whole mount separate from the body, has about 30 conspicuous papillae along its margin, much like those illustrated by Moore (1901).

Actinobdella triannulata, the most common species in the genus, seems to have a predilection for the suckers: *Catostomus fecundus, C. catostomus, C. commersoni,* and *C. macrocheilus* (Hoffman, 1967). The species has been reported from British Columbia (Bangham and Adams, 1954), Wyoming (Bangham, 1951), Lake Huron (Bangham, 1955), and Ontario (Meyer and Moore, 1954). That *A. triannulata* and *Placobdella pediculata* are the only two American glossiphoniids with a strong partiality for fish may reflect a systematic as well as an ecological relationship.

GENUS *OLIGOBDELLA* MOORE, 1918

This genus, originally assigned the preoccupied name *Microbdella*, is unique in having biannulate rather than triannulate segments, but otherwise it is poorly defined. *Oligobdella* has its nearest affinities in and around Japan and New Zealand, but not enough is known about these forms to make any generalizations.

Oligobdella biannulata (Moore, 1900)

Microbdella biannulata Moore, 1900:50, figs. 1-8; Meyer, 1968:9.
Oligobdella biannulata: Moore, 1918:654, fig. 1000; Autrum, 1936:33; Moore, 1959:550, fig. 23.5; Mann, 1961b:159; Meyer, 1968:26; Sawyer, 1971b:54.

Remarks. This obscure amphibian leech was known, until now, only from the original 70-year-old description. A number of specimens of *O. biannulata* from several localities in the southern Appalachians were recently brought to me and are reported elsewhere (Sawyer, 1971b). In spite of its biannulate condition, to which great taxonomic importance has been given, this species resembles the true amphibian leeches, *Batracobdella*, in other ways: feeding on the salamander *Desmognathus*, the translucent olive-green body color, the single confluent pair of eyes, the seven pairs of crop caeca, and the position of the mouth. The position of the mouth in the cup of the fore sucker rather than on the anterior rim places this species among the Glossiphoninae, along with *Batracobdella*, rather than among the Haementarinae, where it has sometimes been placed (Mann, 1961b). The biannulate condition may have been due to immaturity of the specimens. The finding of mature sperm in the few individuals sectioned by Moore does not prove that the specimens were mature, as shown by Sawyer (1970b) in the marine species *Oceanobdella blennii*. Other characters indicating that they were immature include their small size, relatively large suckers, and faint indications of "incipient subdivision of the major annulus."

Five of Moore's syntypes are on deposit in the U.S. National Museum (USNM No. 36394).

GENUS *OCULOBDELLA* AUTRUM, 1936

The genus *Oculobdella*, which has an anterior mouth, a single pair of well-separated eyes, and snail-eating habits, is known only from two North American species. The type-species, *O. socimulcensis* from Lago de Xochimilco, Mexico, has conspicuous dorsal papillae and gonopores separated by two annuli (Caballero, 1931b), whereas the other, *O. lucida*, has a smooth dorsal surface and united gonopores (Meyer and Moore, 1954). *Oculobdella* is closely related to, and may be congeneric with, the South American genus *Anoculobdella* Weber, 1915, which also has an anterior mouth, apparent absence of eyes, three to five rows of dorsal papillae, and gonopores separated by one annulus. *Anoculobdella* is represented by two little-known species which differ from each other in the number and arrangement of dorsal papillae: *A. brasiliensis* Weber, 1915, from Brazil and *A. tribuberculata* Weber, 1915, from Brazil and Paraguay.

Both *Oculobdella* and *Anoculobdella* have the characteristic anterior position of the mouth of the subfamily Haementeriinae, to which *Placobdella* belongs, but their general appearance, certain internal characters (such as fewer than seven pairs of gastric caeca), snail-eating habits, and attachment of the egg sacs to the ventral surface of the body place them much closer to *Helobdella*, of the subfamily Glossiphoniinae.

Oculobdella lucida Moore, 1954

Oculobdella lucida Meyer and Moore, 1954:68, pls. I-II; Moore, 1959:551; Mann, 1961b:159; Moore, 1964:1; Moore, 1966a:10; Meyer, 1968:9; Sawyer, 1968:228; Scudder and Mann, 1968:208.

Description (Fig. 5E). *Oculobdella lucida* has a single common gonopore, but it is often small and can easily be overlooked. The body is a uniform grayish-blue color on the dorsal and ventral surfaces, on both of which occur thin but distinct dark paramedial lines from just anterior to the anus to the neck region. On the dorsum another, less distinct, pair of lines is situated laterad to this pair. None of the *Helobdella* with which it is most closely allied has a pair of such paramedial lines on the dorsal surface. The background grayish-blue color results from numerous diffusely arranged chromatophores more or less uniformly distributed. Unlike most *Placobdella* and *Batracobdella*, there is no marginal pigment pattern. In some specimens there are small indistinct unpigmented areas or spots, metamerically arranged in four rows just lateral to the two pairs of paramedial lines. The four rows of dots may be lacking in an especially heavily pigmented individual, but otherwise there is little individual variation in the basic pigmentation pattern of this species. In some specimens moderately large unpigmented patches are irregularly situated on the dorsum, but in most cases the internal pair of paramedial lines is unaffected.

Some of the individuals tentatively identified as *O. lucida* in the present study were serially sectioned and others mounted whole in Canada balsam to reveal the following characters, which agree with the original description of that species by Meyer and Moore (1954): anterior position of mouth, large well-separated eyes on the fourth annulus, small unbranched gastric caeca, large intestinal caeca, and apparent absence of posterior crop caeca.

Ecology. Throughout Michigan I examined scores of collecting stations of many diverse types, but *O. lucida* was encountered, often in great numbers, at only a few stations, most notably in temporary or semipermanent, almost stagnant, ponds or streams. This species, which is known to eat snails (Moore, 1964), was usually encountered in association with several species of snails, including *Lymnaea* and *Physa*, but it was never observed feeding on them.

Known Distribution (Fig. 31). *Oculobdella lucida*, a poorly known species encountered in the present study in only three localities in Michigan, has previously been reported from British Columbia (Scudder and Mann, 1968), Alberta (Moore, 1964), Manitoba (Meyer and Moore, 1954), and Michigan (Sawyer, 1967). This is only the second record of its occurrence in the United States.

Family Erpobdellidae

GENUS *ERPOBDELLA* BLAINVILLE, 1818

Erpobdella has unsubdivided annuli of equal width and a pre-atrial loop of the vas deferens extending to ganglion XI. For many years the genus *Dina*, which has a slight subdivision of every fifth annulus, was considered a subgenus of *Erpobdella*, and this problem remains unsettled. Similarly, some authors (e.g., Pawlowski, 1955) have considered *Mooreobdella*, a natural group of American species, as a subgenus of *Erpobdella*, but its generic standing is now generally accepted.

Although Europe and Asia have at least four species of *Erpobdella*, only *E. punctata* occurs in North America north of Mexico. It is related to *E. octoculata* of Europe, Africa, and Asia and is one of the most commonly encountered and widely distributed leeches in North America.

Erpobdella punctata (Leidy, 1870)

?*Nephelis quadristriata:* (not Grube, 1851:110); Verrill, 1872b:133; Verrill, 1874a:675; Verrill, 1874b:623; Verrill, 1875a:960; Ward, 1902:276; Moore, 1952:3.
Nephelis punctata Leidy, 1870:89; Moore, 1952:3.
Nephelis lateralis: Verrill, 1872b:133; Nicholson, 1873:493; Verrill, 1874a:675; Verrill, 1874b:623; Bristol, 1897:35; Bristol, 1898:17, text: figs. 1-3, plate: figs. 2-20; Graf, 1899:223; Moore, 1952:3.
Nephelis marmorata: Verrill, 1872b:134; Verrill, 1874a:676; Moore, 1952:3.
?*Nephelis vermiformis:* Nicholson, 1873:493.
?*Nephelis 4-striata:* Forbes, 1893:218.
Herpobdella punctata: Moore, 1898:559; Moore, 1918:659, fig. 1008; Kraatz, 1921:150; Miller, 1929:10; Meyer, 1937a:250; Meyer, 1937b:118; Miller, 1937:85; Richardson, 1942:67.
Erpobdella punctata: Moore, 1901:532; Ward, 1902:276; Moore, 1906:157, fig. 1; Hankinson, 1908:232; Moore, 1912:121, fig. 39; Ryerson, 1915:166; Hankinson, 1916:118; Muttkowski, 1918:391; Moore, 1920:90; Moore, 1922:8; Moore, 1923:15, pl. 1F; Baker, 1924:109; Moore, 1924b:28; Richardson, 1925a:348; Richardson, 1925b:415; Mullin, 1926a:37, pl. V, figs. 1-2; Bere, 1929:177; Bere, 1931:440; Moore, 1936:113; Townes, 1937:167; Richardson, 1943:90; Mathers, 1948:397, pls. 3-4; Pawlowski, 1948:336; Kenk, 1949:38; Moore and Meyer, 1951:69; Pennak, 1953:315; Beck, 1954:74; Meyer and Moore, 1954:92; Oliver, 1958:163; Moore, 1959:556, fig. 23.12; Paloumpis and Starrett, 1960:416; Moore, 1964:2; Moore, 1966a:11; Patrick *et al.*, 1966:342; Thomas, 1966:202; Sawyer, 1967:36; Carlson, 1968:164; Judd, 1968:7; Sapkarev, 1968:226; Sawyer, 1968:228; Scudder and Mann, 1968:208; Judd, 1969:168; Meyer, 1969:161; Herrmann, 1970:5; Sawyer, 1970a:85.
Erpobdella punctata annulata: Moore, 1922:8; Bere, 1931:440; Moore and Meyer, 1951:69; Meyer, 1968:24.
?*Dina fervida:* Miller, 1929:34.

Dina lateralis: Moore, 1952:3; Moore, 1959:556; Mann, 1961b:168.
Erpobdella lateralis: Moore, 1952:3.
Erpobdella puctata: Mann, 1961b:168.
Erpobdella annulata: Mason et al., 1970:R323.

Description (Fig. 8C). *Erpobdella punctata* has a paramedial (and sometimes also a paramarginal) pair of black pigment concentrations, which form two (or four) conspicuous longitudinal stripes. On the dorsum (and less conspicuously on the venter) of every annulus occur 8-16 small white-tipped papillae, especially noticeable in heavily pigmented individuals. The middorsal line and the entire venter are usually unpigmented, but the degree of pigmentation varies from heavily pigmented, almost black individuals in which the longitudinal stripes are almost obscured to almost cream-colored individuals in which the paramedial stripes are suggested only by a few pigment concentrations (Fig. 8D).

The gonopores are invariably located in the furrows and are separated by two annuli (Fig. 9F). The male gonopore of fully mature individuals is especially large and glandular. The atrium is correspondingly large and well developed, and the atrial cornua of fully mature individuals are slightly coiled, closely resembling the condition found in *Nephelopsis obscura*. On the other hand, the male gonopore and atrium of immature individuals are so small and poorly developed (Figs. 11B, I) that unless one had examined hundreds of individuals, one would hesitate to call them the same species.

Ecology. Various aspects of the life history, fecundity, behavior, and feeding habits of *Erpobdella* have been investigated (Sawyer, 1970a). The life cycle may require one year or two, depending upon local conditions, but in either case few seem to survive to a second breeding season. There is evidence of mass movements upstream in early spring in some localities. During April and May in Michigan each individual lays on the average about ten cocoons, each containing about five eggs which hatch in three to four weeks. The presence of its large (8-9 mm) distinctive cocoons (Fig. 10E) in a pond or stream is usually the first, and sometimes the only, indication that the species is there. Intraspecific and snail predation of cocoons contributes markedly to the high mortality rate, about 93 percent during the first year (Sawyer, 1970a). Like all American erpobdellids, *E. punctata* is a scavenger and predator rather than a parasite, aquatic insects and oligochaetes constituting the major part of the diet. It can serve as host for juvenile nematomorphs (Sawyer, 1971).

Distribution (Fig. 30). *Erpobdella punctata* is one of the most common and widely distributed of the North American leeches. It is especially abundant in the northern United States and Canada, but in the southern United States it appears to be replaced as the dominant erpobdellid by *Mooreobdella microstoma*. Published records believed reliable include Mexico (Caballero, 1941), Alaska (Meyer and Moore, 1954), British Columbia (Clemens et al., 1939; Scudder and Mann, 1968), Alberta (Bere, 1929; Moore, 1964), Saskatchewan (Moore and Meyer, 1951; Oliver, 1958), Ontario (Moore, 1906, 1924b, 1936; Faull, 1913; Ryerson, 1915; Meyer and Moore, 1954; Thomas, 1966; Judd, 1968, 1969), Manitoba, Quebec (Meyer and Moore, 1954), New Brunswick (Richardson, 1943), Newfoundland, St. Pierre, Miquelon, and Nova Scotia (Pawlowski, 1948; Gates and Moore, 1970), California (Verrill, 1875a; Hagadorn, 1958), Oregon (Mason et al., 1970), Utah (Beck, 1954), Colorado (Verrill, 1874b, 1875a; Herrmann, 1970), Wyoming (Verrill, 1874a; Moore, 1898), Nebraska (Verrill, 1874a), Iowa (Mathers, 1948; Carlson, 1968), Minnesota (Moore, 1912), Wisconsin (Muttkowski, 1918; Baker, 1924; Bere, 1931; Miller, 1937; Sapkarev, 1968), Michigan (Hankinson, 1908, 1916; Miller, 1937; Kenk, 1949; Sawyer, 1968, 1970a), Illinois (Moore, 1901; Baker, 1922; Richardson, 1925; Paloumpis and Starrett, 1960), Indiana (Moore, 1920), Ohio (Moore, 1906; Kraatz, 1921; Miller, 1929), Pennsylvania (Verrill, 1874a; Moore, 1906, 1912), New York (Moore, 1923; Barrow, 1953), New Jersey (Verrill, 1874a), Connecticut (Verrill, 1874a; Barrow, 1953), Massachusetts (Verrill, 1874a; Weston and Turner, 1917), Missouri (Meyer, 1937a), Louisiana (Sawyer, 1967), South Carolina (Sawyer, 1971c), and Georgia (Patrick et al., 1966). It has never been previously reported from South Carolina, Delaware, and Maryland.

GENUS *NEPHELOPSIS* VERRILL, 1872

This genus, represented only by *N. obscura*, has a conspicuous subdivision of most of the mid-body annuli. Its phylogenetic affinities are obscure, but several authors (e.g., Moore and Meyer, 1951) have suggested a close relationship with the European genus *Trocheta* Dutrochet, 1817, which has even more subdivisions of the mid-body annuli. Whether *Trocheta* should be revised to incorporate the monotypic *Nephelopsis* will depend upon a detailed morphological and biological comparison of the two genera.

Nephelopsis obscura Verrill, 1872

Nephelopsis obscura Verrill, 1872b:135; Verrill, 1874a:674; Verrill, 1874b: 623; Verrill, 1875a:958; Ruthven, 1906:51; Moore, 1912:123, figs. 35-36, 40; Ryerson, 1915:166; Moore, 1918:659, fig. 1009; ?Muttkowski, 1918: 391; Moore, 1922:8; Baker, 1924:109; Moore, 1924b:29; Mullin, 1926a:37;

Bere, 1929:177; Rawson, 1930:36; Bere, 1931:440; Meyer, 1937b:119; Richardson, 1942:67; Richardson, 1943:90; Mathers, 1948:397, pl. 4; Pawlowski, 1948:338, fig. 5; Meyer and Bangham, 1950:20; Moore and Meyer, 1951:70; Moore, 1952:3; Pennak, 1953:315, fig. 201A; Beck, 1954: 74; Meyer and Moore, 1954:93; Oliver, 1958:163; Moore, 1959:555, fig. 23.11; Mann, 1961b:168, fig. 23; Moore, 1964:2; Moore, 1966a:11; Thomas, 1966:202; Sapkarev, 1968:226; Scudder and Mann, 1968:208; Herrmann, 1970:5.
?*Nephelis obscura* var. *maculata:* Forbes, 1893:216.
?*Nephelis maculata:* Forbes, 1893:216.

Description (Fig. 8H). *Nephelopsis obscura* is moderately large (5-6 cm or larger) and is light brown with irregularly scattered black splotches dorsally and, less commonly, ventrally. It has a two-annuli separation of the gonopores, which are usually in the furrows, but the female gonopore may be slightly more caudad (Figs. 9G, 11K). In earlier keys emphasis was placed on the significance of the spirally coiled atrial cornua for distinguishing *N. obscura* from related erpobdellids, but in practice it is exceedingly difficult to distinguish it from *Erpobdella* and *Dina* on this character alone. There is a pre-atrial loop of the vas deferens which extends to ganglion XI. Most of the mid-body annuli are conspicuously subdivided, but this character alone is not reliable for separating it from poorly preserved, wrinkled specimens of *Dina* or *Erpobdella*.

Ecology. Analyses of stomach contents of *N. obscura* have revealed large numbers of insect larvae, which are probably its basic diet (Forbes, 1893; Moore, 1912; Moore, 1966a), but it has also been reported to feed upon oligochaetes, snails, dead fish, and wastes from a fish-packing station (Moore, 1924). In the laboratory I observed an unfed *N. obscura* eating an immature individual of *E. punctata*. Meyer and Bangham (1950) found it in the air bladder of a lake trout (*Salvelinus namaycush*). The large, uniquely shaped cocoons (Fig. 10F) were found by myself and others on 2 and 14 May, 15 July, 2 August, and 7 October, which suggests a summer breeding season. The cocoons were briefly described by Verrill (1875a), who stated that each one contained five to ten eggs or young.

Distribution (Fig. 31). *Nephelopsis obscura* appears to be restricted to the cold waters of the extreme northern United States, the Rocky Mountain region, and Canada. Published records believed valid include British Columbia (Clemens *et al.*, 1939; Scudder and Mann, 1968), Alberta (Bere, 1929; Moore, 1964), Saskatchewan (Moore and Meyer, 1951; Oliver, 1958), Ontario (Ryerson, 1915; Moore, 1924; Thomas, 1966), Utah (Beck, 1954), Wyoming (Forbes, 1893), Colorado (Verrill, 1874b, 1875a; Herrmann, 1970), Minnesota (Moore, 1912), Iowa

(Mathers, 1948), Wisconsin (Verrill, 1874a; Muttkowski, 1918; Baker, 1924; Bere, 1931; Pawlowski, 1948; Sapkarev, 1968), and Michigan (Ruthven, 1906; Adams, 1908). It has not been found south of Iowa, Wisconsin, Michigan, or Ontario.

GENUS *DINA* R. BLANCHARD, 1892

Various authors have considered *Dina* a subgenus of *Erpobdella*, with which it shares the presence of a pre-atrial loop of the vas deferens to ganglion XI, but it differs in having every fifth annulus widened and subdivided. In practice it is usually difficult to distinguish such subdivision in *Dina* from winkled, preserved *Erpobdella* or *Mooreobdella*. Several species of *Dina* live in Europe and Asia, and at least two species, *D. dubia* and *D. parva*, occur in the midwestern United States. An obscure species, the eyeless *D. anoculata*, from California has barely been mentioned in the literature since the original description by Moore (1898). The poorly defined *D. lateralis* is probably a synonym for *Erpobdella punctata*.

Dina parva Moore, 1912

Dina parva Moore, 1912:125, figs. 33-34, 41, 43; Moore, 1918:659; Moore, 1920:90; Moore, 1922:8; Baker, 1924:109; Moore, 1924b:30; Richardson, 1925a:373; Mullin, 1926a:37; Bere, 1929:177; Rawson, 1930:36; Bere, 1931:440; Moore, 1936:114; Mathers, 1948:397; Moore, 1949:38; Pennak, 1953:316, figs. 200C, D; Meyer and Moore, 1954:92; Oliver, 1958:163; Moore, 1959:556; Mann, 1961b:168; Moore, 1964:2; Moore, 1966a:11; Thomas, 1966:202; Meyer, 1968:24; Sapkarev, 1968:226; Herrmann, 1970:5; Mason et al., 1970:R323.

Description (Fig. 8B). This poorly known species is the only American erpobdellid with the following combination of characters: a pre-atrial loop of the vas deferens extending to ganglion XI, every fifth annulus widened and subdivided, gonopores usually separated by three and a half annuli, and the typical smoky-blue coloration of other unpigmented erpobdellids. The arrangement of the gonopores is similar to that of *D. dubia*, the male gonopore being on the ring and the female being in the furrow three and a half annuli caudad (Fig. 9B). Moore (1922, 1936) has reported individuals with only two and a half and three annuli between the gonopores respectively. The large male bursa, which in the few specimens examined appears to be protruded cephalad much more often than in *D. dubia*, is cylindrical with a flattened tip.

Dina parva is so closely related to the conspicuously pigmented *D. dubia* that the possibility that they represent two forms of the same species needs to be investigated. Earlier keys distinguished them on the basis of size, *D. parva* supposedly never larger than 2 cm and *D. dubia* usually about 4-5 cm, but in Hammond Bay, Michigan, an individual of *D. parva*, 4.2 cm long, was found which was otherwise identical to the smaller individuals of the same population.

Ecology. Almost nothing is known about reproduction and feeding of this species. Mathers (1948) reported it to be predacious, feeding upon aquatic insect larvae, snails, and worms, and Moore (1920) reported it to be a scavenger, feeding upon dead turtles and a dead shrew.

Distribution (Fig. 32). *Dina parva* is a relatively uncommon species of the Great Lakes region which, like *D. dubia*, has a northern distribution. Published records believed reliable include Alberta (Bere, 1929), Saskatchewan (Oliver, 1958), Lake Simcoe, Ontario (Rawson, 1930), Lake Nipigon (Moore, 1924b), Lake Nipissing (Moore, 1936), Lake Superior (Thomas, 1966), Colorado (Herrmann, 1970), Iowa (Mathers, 1948), Minnesota (Moore, 1912), Oregon (Mason *et al.*, 1970), Wisconsin (Baker, 1924; Bere, 1931; Sapkarev, 1968), Indiana (Moore, 1920), and Illinois (Richardson, 1925a).

Dina dubia Moore and Meyer, 1951

Dina dubia Moore and Meyer, 1951:70; Beck, 1954:74; Meyer and Moore, 1954:66; Oliver, 1958:163; Moore, 1959:556; Mann, 1961b:168; Mathers, 1963:173; Moore, 1964:13; Moore, 1966a:11; Meyer, 1968:23; Herrmann, 1970:5.

Description (Fig. 8A). This moderate-sized species, 2.5-5 cm, is the only American erpobdellid with the following combination of characters: a pre-atrial loop of the vas deferens, gonopores separated by three and a half or four annuli (Fig. 9A), and a heavily mottled dorsum, usually with a middorsal stripe. The dark greenish dorsum of most individuals is heavily mottled with a black pigment, especially along the middorsal line, to create a characteristic dark middorsal stripe. However, a number of individuals can often be found even within a population with the stripe missing or poorly developed.

Ecology. I encountered a breeding population on 8 May 1967 in southern Michigan (water 12°C). Individuals isolated in separate laboratory containers laid an average of 7.9 (5-13) cocoons per individual (Fig. 12D), each cocoon containing on the average 4.15 (1-9) eggs (Fig. 12E). Like the other American erpobdellids studied, the rate of cocoon-laying decreased rapidly after the initial outburst. Little is known about its diet, except that insect larvae were found in the guts of dissected individuals.

Distribution (Fig. 32). *Dina dubia* appears to be a northern species distributed from the Great Lakes west to Alaska. Published records believed reliable include only Alaska (Moore and Meyer, 1951), Alberta (Moore, 1964), Saskatchewan (Oliver, 1958), Colorado (Herrmann, 1970), Utah (Beck, 1954), and Iowa (Mathers, 1963). It has never been previously reported from Missouri, Illinois, and Michigan.

GENUS *MOOREOBDELLA* PAWLOWSKI, 1955

On the basis of their lack of a pre-atrial loop of the vas deferens extending to ganglion XI, Pawlowski (1955) removed *Dina fervida*, *D. microstoma*, and *D. bucera* from the genus *Dina* and placed them in *Mooreobdella*, a new subgenus of *Erpobdella*.

Moore (1959) later recognized *Mooreobdella* as a distinct genus, whereas other authors (Soós, 1963, 1966) considered *Mooreobdella* to be a subgenus of *Dina*, the systematic position of which is still unsettled.

Only three species of *Mooreobdella* have been described, *M. fervida* (Verrill, 1874a), *M. microstoma* (Moore, 1901), and *M. bucera* (Moore, 1949), none of which has been found as far south as Mexico. Other forms of *Mooreobdella* will probably be discovered eventually, most likely among the poorly known leech fauna of the southern states.

Mooreobdella fervida (Smith and Verrill, 1871)

Nephelis fervida Smith and Verrill, 1871:451; Verrill, 1872b:134; Verrill, 1874a:676; Moore, 1952:3; Moore, 1953:11.

Dina fervida: Moore, 1901:535, fig. 36; Moore, 1906:157, fig. 6; Hankinson, 1908:232; Moore, 1912:127, fig. 42; Hankinson, 1916:118; Moore, 1918: 660, fig. 1010; Moore, 1920:90; Moore, 1922:8; Mullin, 1926a:37; ?Miller, 1929:34; Moore, 1936:114; ?Meyer, 1937a:250; Meyer, 1937b:119; Miller, 1937:85; Richardson, 1942:68; Richardson, 1943:89; Mathers, 1948: 397, pl. 3; Pawlowski, 1948:318; Moore, 1949:38; Moore and Meyer, 1951:73; Moore, 1952:3; Moore, 1953:11; Pennak, 1953:316, fig. 201E; Meyer and Moore, 1954:92; Oliver, 1958:163; Herrmann, 1970:5.

Dana fervida: Mathers, 1948:412.

Mooreobdella fervida: Moore, 1959:555, fig. 23.13; Fredeen and Shemanchuk, 1960:733; Mann, 1961b:167; Moore, 1964:2; Moore, 1966a:11; Sawyer, 1967:35; Herrmann, 1970:2.

Description (Fig. 8G). This small (2-4 cm) smoky-gray species is usually without any black pigmentation. An occasional individual or even a population, however, is encountered with numerous minute black chromatophores scattered throughout the dorsal and ventral surfaces. The gonopores are separated by two annuli (Fig. 9E), usually in the furrows. *Mooreobdella fervida* most closely resembles *M. bucera*, with which it may someday prove conspecific, but differs somewhat in the arrangement of the atrial horns, those of *M. fervida* projecting more cephalad.

Ecology. Apart from a few cocoons of *M. fervida* encountered on 27 May (water 23.5°C) in southwestern Michigan, almost nothing is known about its breeding habits.

Moore (1912, 1920) reported that its stomach contents included mainly tubificid worms and some insect larvae and that it is commonly found on dead animals (coot, duck, shrew) at the water's edge.

Distribution (Fig. 33). *Mooreobdella fervida,* represented in the present study from a few localities in Illinois, Michigan, and Minnesota, is a northern species which occurs over much of Canada and the northern United States. It is not known from the western and northeastern states nor any farther south than extreme northern Illinois, Indiana, Ohio, and Pennsylvania, where it becomes replaced by the closely related southern species *M. microstoma.* Published records thought to be valid include Alberta (Moore, 1964), Saskatchewan (Oliver, 1958), Ontario (Moore, 1906, 1936), Nova Scotia (Meyer and Moore, 1954), Prince Edward Island, New Brunswick (Richardson, 1943), Colorado (Herrmann, 1970), Minnesota (Moore, 1912), Iowa (Mathers, 1948), Michigan (Hankinson, 1908, 1916; Miller, 1937), Illinois (Moore, 1901), Indiana (Moore, 1920), Ohio (Moore, 1906; Miller, 1929), and Pennsylvania (Moore, 1906).

Mooreobdella microstoma (Moore, 1901)

Dina microstoma Moore, 1901:587, fig. 37; Moore, 1906:157; Moore, 1912:128; Moore, 1918:659; Richardson, 1928:407; Miller, 1929:10; Moore, 1936:114; Meyer, 1937a:250; Miller, 1937:90; ?Kenk, 1949:38; Moore and Meyer, 1951:73; Pennak, 1953:316; Meyer and Moore, 1954:66; Meyer, 1968:23; Herrmann, 1970:5.

Mooreobdella microstoma: Moore, 1959:555; Mann, 1961b:167; Sawyer, 1967:34; Meyer, 1968:26.

Description (Fig. 8E). This small (3-5 cm) smoky-gray (unpigmented) species can always be distinguished from its congenitors, *M. fervida* and *M. bucera,* by having a three-annuli separation of the gonopores rather than two annuli (Fig. 9C). They usually lie in the furrows rather than on the rings, but an occasional individual can be found with the gonopores slightly upon the rings (Fig. 11H). The atrial horns usually project laterad at right angles to the body axis, whereas the horns of both *M. fervida* and *M. bucera* project somewhat cephalad (Figs. 9C-E). Occasionally an individual of *M. microstoma* is found, however, in which one or both of the horns project slightly cephalad (Fig. 11A).

Ecology. Numerous cocoons (Fig. 10A) of *M. microstoma* were encountered on 1 June 1967 in the Raisin River in southeastern Michigan (water 19°C), but otherwise almost nothing is known about its breeding habits. The species is a host for juvenile nematomorphs (Sawyer, 1971c).

Our knowledge of its feeding habits consists of the brief report by Miller (1929) of finding small tubificid worms in the stomachs of the few individuals examined.

Distribution (Fig. 33). *Mooreobdella microstoma* is especially abundant in the southern states, extending up the Mississippi and Ohio River systems and into Lake Erie. It occurs as far north as extreme northern Illinois and Indiana and barely gets into extreme southeastern Michigan and the southern shores of Lake Erie, where it slightly overlaps the range of the closely allied northern form *M. fervida*. The occurrence of *M. microstoma* north of Illinois and Lake Erie and in the western and northeastern states has not been confirmed, in spite of one or two reports to the contrary (Gee, 1913; Moore, 1936). Although it has been found in southeastern Texas, it has not yet been found in Mexico.

Published records thought to be valid include Illinois (Moore, 1901; Richardson, 1928), southeastern Michigan (Kenk, 1949), Ohio (Moore, 1906; Miller, 1929), northwestern Pennsylvania (Moore, 1906), Missouri (Meyer, 1937a), Colorado (Herrmann, 1970), Louisiana (Sawyer, 1967), and South Carolina (Sawyer, 1971c). It has not been previously reported from Kansas and Texas.

Mooreobdella bucera (Moore, 1949)

Dina bucera Moore, 1949:38; Moore and Meyer, 1951:73; Moore, 1953:9, fig. 2, pl. 1; Meyer, 1968:23.
Mooreobdella bucera: Moore, 1959:555; Mann, 1961b:167; Sawyer, 1967:35; Meyer, 1968:26; Sawyer, 1968:228.

Description (Fig. 8F). *Mooreobdella bucera*, which is a small (2-3 cm) smoky-gray species without any black pigmentation, has the gonopores separated by two annuli (Fig. 9D). The atrial horns of *M. bucera* may in some ways be considered intermediate between *M. microstoma* and *M. fervida*, but it most closely resembles the latter. The circumesophageal nerve commissure appears as a characteristic white band around the neck. Earlier keys (Moore, 1959; Mann, 1961b) erroneously separated *M. bucera* and *M. fervida* on the basis of the relative position of the gonopores. Although some populations of *M. bucera* have the gonopores primarily in the furrows (Figs. 11C, D), an examination of 51 individuals from Earhardt Pond, Michigan, located only a few miles from the type-locality of *M. bucera*, corroborated Moore's (1953) observation that most had the gonopores on the

TABLE 1

SUMMARY OF OBSERVATIONS ON A POPULATION OF *M. BUCERA*

Observations were made at Earhardt Pond, Washtenaw County, Michigan, during the spring of 1967. The numbers and weights refer to the individuals collected at monthly intervals in a one-hour period starting one hour after sunset.

Date	N	Mean Weight (mg)	Range (mg)	Water Temperature (°C)
5 March	0	Iced over		0.3
5 April	0	Partially iced over		15.0
9 April	..	First appearance of *M. bucera*		13.0
16 April	..	First cocoons		15.0
6 May	51	84.9	50–140	15.0
3 June	19	74.2	25–105	18.0
6 July	0	..	..	23.5

rings, the position of the female pore varying slightly (Figs. 11E-G). There is a need for a detailed comparison of *M. bucera* and *M. fervida*.

Ecology. The little evidence available suggests that *M. bucera* may appear in the spring somewhat later than *Erpobdella punctata*. On the nights of 29 March (water 3.5°C), 30 March (5.5°C), and 1 April (7.5°C) both species had begun to deposit cocoons (Table 1). On 10 April individuals of *M. bucera* from Earhardt Pond were isolated in individual laboratory containers. On the average each individual laid 4.6 cocoons (Fig. 12B), each of which contained an average of 5.95 eggs (Fig. 12C). The cocoons deposited initially contained an average of over 7.1 eggs, after which the number of eggs decreased rapidly to 6.6 on the second day and to 5.6 on the fifth day (Fig. 12A). Similarly, the number of cocoons deposited per individual was greatest initially, after which the number rapidly declined.

The process of cocoon deposition in *M. bucera* was observed a number of times and closely resembles that described for *Erpobdella punctata* by Sawyer (1970a). The clitellar secretions form an elastic band or incipient cocoon which adheres to the substrate. After loosening the cocoon from the skin by several rotations of the body, the body slips posteriorly, leaving the flaccid cocoon attached to the substrate in its initial position. The cocoon is shaped somewhat by the anterior sucker and is later ventilated by undulating movements of the body. There was no opportunity to observe whether this ventilation would continue if the cocoon were removed, as it does in *Erpobdella punctata*. One apparent difference noted between *M. bucera* and *E. punctata* is that no intraspecific predation of cocoons was ever observed in *M. bucera*, despite numerous opportunities on the part of the leeches to eat the newly laid cocoons. Such intraspecific predation of cocoons was commonly observed in *E. punctata* under essentially the same laboratory conditions (Sawyer, 1970a).

Mooreobdella bucera seems to have a preference for small temporary and permanent ponds without drainage and is not usually found in lakes, rivers, or even large streams (Kenk, 1949; Sawyer, 1968). This is in apparent contrast to its congenitors, *M. fervida* and *M. microstoma*, both of which abound in lakes and rivers. Little is known about the feeding habits of *M. bucera*, but, like the other American erpobdellids, it is probably a scavenger and predator rather than a parasite.

At monthly intervals from March to July 1967 a population of *M. bucera* was investigated at Earhardt Pond (Table 1). The peak number of individuals occurred from mid-April to early May, after which there was a decline. No adult *M. bucera* were found after mid-June. It is probable that they died after the breeding season because in the laboratory all the *M. bucera* died after breeding, unlike *Erpobdella punctata*, which were kept under essentially the same conditions. In order to get a picture of the population structure and life cycle of *M. bucera*, the 51 individuals collected on 6 May, just after the peak of the cocoon-laying period, were weighed, using a method similar to that of Mann (1953). Judged from the size of the gonopores and condition of the clitellum, all of the individuals were sexually mature. The distribution of the individual weights was unimodal, with a mean at 84.9 mg and a range of 50-140 mg. A similar collection made a month later on 3 June, toward the end of the breeding season, also had a unimodal distribution, with a mean at 74.2 mg and a range of 25-105 mg. There seemed to be a considerable individual weight loss by the end of the reproductive season, a phenomenon also noted in *Erpobdella punctata* by Sawyer (1970a). The unimodal weight distributions, the absence of juveniles during the breeding season, and the decline in the numbers of adults after the breeding season all suggest a simple annual life cycle for *M. bucera*, at least in the permanent pond studied.

Distribution (Fig. 33). This poorly defined species had previously been reported only from Washtenaw and Livingston counties, Michigan (Kenk, 1949; Sawyer, 1968), the same counties in which it was encountered in the present study. The restricted range of this species lies along the extreme southern part of the range of the closely related form *M. fervida*, near or in the area of apparent overlap of the ranges of the latter and *M. microstoma*.

Family Hirudinidae*

GENUS *HAEMOPIS* SAVIGNY, 1820

The genus *Haemopis* includes the largest and, in some ways, the most interesting of the Holarctic leeches, but little is known about the ecology, morphology, and systematic standing of the various species (Soós, 1969). Europe has only one representative, *H. sanguisuga* (L.), whereas the northern United States and Canada have six described forms. Of these six species, only *H. marmorata, H. grandis*, and *H. terrestris*, by far the most common and widespread, will be discussed below. The other three, more obscure, forms are presented only in the key. The revision of North American *Haemopis* by Richardson (1969), who placed *H. marmorata, H. terrestris, H. lateromaculata*, and *H. kingi* into the new genus *Percymoorensis* and placed *H. grandis* and *H. plumbea* into the monotypic genera *Mollibdella* and *Bdellarogatis* respectively, has not yet been generally accepted by students of the genus. In light of the detailed morphological investigations of *H. marmorata, H. grandis,* and *H. terrestris* presented below, I cannot support the proposed revision in this study.

The genus can be separated into two groups on the basis of the degree of subdivision of the VIIa3 and VIIIa1 annuli, those of *H. terrestris* being completely subdivided and those of all other species being undivided or only faintly subdivided dorsally. The systematic significance of this useful character is still unsettled. Without a large sample of several species of *Haemopis* it may be difficult to judge with certainty whether the VIIa3 and VIIIa1 annuli are, indeed, subdivided. In the absence of distinct metameric guidelines, counting the dorsal annuli is useless because the dorsal parts of these annuli may have shallow furrows suggestive of incipient subdivisions, the ventral parts remaining distinctly undivided. A more reliable method is to count only the ventral annuli from the oral cavity to the male gonopore, but even in this method one should not be confused by the varying degrees of subdivisions of the ventral aspects of the first two annuli (V and VIa1) behind the oral cavity. If one lets the annulus bearing the fifth pair of eyes be the second ventral annulus, *H. terrestris* has 27 distinct annuli from the oral cavity to the annulus (XIb6) bearing the male gonopore, whereas *H. marmorata, H. grandis,* and apparently all other species of *Haemopis* have only 25 annuli (Figs. 14D-F).

---

*The widely accepted familial name Hirudinidae is based etymologically on the stem of *hirudinis,* the Latin genitive singular of *hirudo,* and should always be used in preference to Hirudidae (Art. 29a of International Code). See Richardson, 1969; Soós, 1969.

The presence or absence of jaws and teeth is another useful character in this genus, but until its true systematic significance is settled, too much reliance on this single character may lead to an unnatural classification. Shortened posterior crop caeca may correlate with the absence of jaws and teeth.

Reliable identifications of the various species of *Haemopis* require careful dissection and cannot be based solely on such external characters as color and pigmentation, except possibly for the distinct *H. terrestris*. The degree of mottling in the two most common species, *H. marmorata* and *H. grandis*, is especially subject to so much inter- and intrapopulation variation that misidentifications can easily occur. My confusion over the identifications of living individuals of *H. marmorata* and *H. grandis*, even within the same population, was removed only after dissections were made (Fig. 15).

Haemopis marmorata (Say, 1824) Moore, 1901

Hirudo marmoratis Say, 1824:266.
Democedes maculatus: Kinberg, 1867:356; Verrill, 1872b:137; Verrill, 1874a: 671; Moore, 1952:3.
Aulastomum lacustris: Leidy, 1868:229.
Aulastomum lacustre: Verrill, 1872b:135; Verrill, 1874a:670; Verrill, 1874b: 623; Verrill, 1875a:958; Pawlowski, 1948:335; Moore, 1952:3; Beck, 1954:73.
Hexabdella depressa: Verrill, 1872b:136; Verrill, 1874a:673; Moore, 1952:3.
Aulostoma lacustris: Forbes, 1893:218.
?*Haemopis sanguisuga:* Blanchard, 1896b:3.
Haemopis marmoratis: ?Moore, 1898:560; Moore, 1901:519, figs. 7, 24, 26, 33-34; Ward, 1902:275; Moore, 1912:110, fig. 32; Cahn, 1915:123; Ryerson, 1915:166; Hankinson, 1916:118; Moore, 1918:658, figs. 1006-1007; Moore, 1920:94; Moore, 1922:8; Moore, 1923:15; Moore, 1924b:28; Miller, 1929:10, fig. 8; Rawson, 1930:35; Bere, 1931:439; Miller, 1933:343; Moore, 1936:114; Meyer, 1937a:250; Meyer, 1937b:118; Miller, 1937:85; Richardson, 1942:68; Miller, 1943:198, figs. 1-3; Richardson, 1943:89; Miller, 1944, 43:177, figs. 1-6; Miller, 1944, 44:31, figs. 1-6; Miller, 1945:233, pls. 1-2; Mathers, 1948:397, pl. 2; Pawlowski, 1948:333, fig. 4; Moore and Meyer, 1951:68; Pennak, 1953:317, fig. 201H; Mathers, 1954:460; Meyer and Moore, 1954:91; Mathers, 1963:168; Lynch et al., 1968:310; Scudder and Mann, 1968:208.
Haemopis marmoratus: Hankinson, 1908:232; Moore, 1920:90.
Haemopis marmorata: Moore, 1923:38, pl. 1; Bere, 1929:177; Moore, 1952:3; Beck, 1954:73; Oliver, 1958:163; Moore, 1959:554, fig. 23.10; Mann, 1961b:164; Moore, 1964:2; Moore, 1966a:11; Sapkarev, 1968:226; Clifford, 1969:583; Herrmann, 1970:5.
Haemopsis marmoratis: Mullin, 1926a:36, pl. III, fig. 1; Miller, 1942:45, figs. 1-4.
Percymoorensis marmoratis: Richardson, 1969:123, figs. 4B, 6B.
Haemopis marmorate: Gates and Moore, 1970:45.

TABLE 2

VARIATIONS OF REPRODUCTIVE AND DIGESTIVE SYSTEMS OF *H. MARMORATA*
(14 individuals from 4 states)

Length (cm)	6.8	4.4	6.2	7.9	5.4	5.1
Origin	Missouri	Delaware	Illinois	Illinois	Michigan	Michigan
Color	Light	Light	Light	Typical	Typical	Typical
Posterior Extension:						
penis sheath	XVI¾	XVI½	XVI½	XVI¾	XVII¾	XVIII¼
vaginal stalk	XVI½	XVII	XV½	XV¼	XVI½	XVI½
crop caeca	XXIII	XXII½	XXIII	XXII¼	XXII	XXII½
Ratio of Short and Long Arms of Penis Sheath	1:2.0	1:2.0	1:1.9	1:1.5	1:2.0	1:1.7
Position:						
center of ovary	XIV	XIV½	XIII½	XIII¼	XIV½	XIII½
anterior edge of prostate gland	XIV	XIII¾	XIII¾	XIII¼	XIV¼	XIII½
sperm sacs, left	XII–XIII½	XII–XIV	XII–XIII½	XI¼–XIII	XI¾–XIV	XI¼–XIV
sperm sacs, right	XII½–XIII	XII¼–XIV¾	XI¾–XII¼	XII¼–XIV	XII–XIII½	XII–XV
Relative to Nerve Cord:						
male opening	Right	Left	Left	Right	Left	Right
female opening	Right	Left	Left	Right	Left	Right
penis sheath	Right	Left	Left	Right	Left	Right
vaginal stalk	Right	Left	Left	Left	Left	Right

Description (Fig. 13D). After dissecting over two dozen individuals of *H. marmorata* from a number of localities (Sawyer, 1969), two distinct but variable color forms were found which could not be separated on the basis of the morphology of the reproductive and digestive systems (Table 2). The typical form has a heavy mottling which covers both the dorsal and ventral surfaces, including the posterior sucker. This mottling consists of dense, closely confluent black, green, and a few yellowish chromatophores, which make the animal appear dark green or even almost black from a distance. Light cream-colored patches appear wherever the chromatophores are sparse or missing. In heavily pigmented individuals small whitish dots representing metameric sensillae are the only signs of metameric pigment patterns, even in the young, which have the same mottled appearance as the adult.

The other color variant, which lacks the heavy mottling of the typical form and superficially resembles *H. grandis,* with which it was often erroneously identified on external examination alone, is a light slate-gray color with a few small black blotches scattered dorsally and ventrally. The occurrence of both varieties side by side in Cook County, Illinois, without any intermediate forms suggests the possibility that the heavily mottled and light forms may be sibling species, but whether these pigmentation differences are combined with other differences of enough systematic importance to justify such a view needs to be investigated.

7.5	4.5	6.3	7.3	7.2	5.8	5.9	5.0
Michigan	Michigan	Michigan	Michigan	Michigan	Michigan	Michigan	Michigan
Typical	Typical	Typical	Typical	Typical	Typical	Typical	Typical
XVIII	XVI	XVII	XVII	XVII½	XVII¾	XVI¾	XVI½
XVII	XV	XVI	XV	XV¾	XVI	XV¾	XV¾
XXII½	l.XXII r.XXIII½	XXII½	XXIII	XXIII	XXIII	XXIII½	XXIII½
1:3.1	1:2.0	1:1.7	1:2.0	1:1.4	1:1.8	1:1.9	1:1.5
XIV½	XIII¼	XIII½	XIII	XIV	XIV	XIII	XIV¼
XV½	XIV	XIV	XIV	XIV	XIV½	XIV	XIII¼
XII–XIV¾	XII¼–XIII¾	XII½–XIV	XII–XIV	XI–XIII	XII¾–XIV¼	XII–XIII	XII–XIII½
XII–XIV	XI½–XIII	XII–XIII½	XI–XIII	XI¾–XIII	XII–XIII¾	XII–XIII½	XII–XIII½
Left	Left	Left	Left	Left	Left	Right	Left
Left	Left	Left	Left	Left	Left	Left	Left
Left	Left	Left	Left	Left	Left	Right	Left
Left	Right	Right	Left	Left	Left	Left	Left

Haemopis marmorata has an exceedingly long and slender penis (Pawlowski, 1948, fig. 4), which is rarely protruded in preserved specimens, unlike that of *H. grandis*. The gonopores, which are separated by five to five and a half annuli, are usually on the anterior third of their respective annuli, but their actual positions are subject to much variation (Fig. 16C). The position of each gonopore varies from the middle of the ring to the next anterior furrow independently of the other gonopore, with the result that a number of intermediate variations are commonly encountered between the extreme conditions in which both gonopores are in the middle of the ring (b6) or in the furrow (b5/b6).

Dissections were made of 14 individuals of *H. marmorata* representing nine localities in Michigan, Illinois, Missouri, and Delaware and ranging in size from 4.4 to 7.9 cm (Sawyer, 1969). Distinctive characteristics of this species are the position of the anterior part of the prostate gland (XIII¼-XV½,* usually XIV), the posterior flexion of the penis sheath (XVI-XVIII¼, usually about XVII), the ratio of the short and long arms of the penis sheath (1:1.4 to 1:2.0, with one individual at 1:3.1, usually 1:2.0), the position of the center of the ovaries (XII-XIV½, usually about XIII¾), the posterior extension

---

*The fractional measurements used in this study refer to the fractional distance between two ganglia; e.g., X¾ means three-quarters of the distance between ganglia X and XI, or halfway between X/XI and XI.

of the vaginal system (XV-XVII, usually about XVI), and the posterior extension of the large posterior crop caeca (XXII-XXIII½, usually XXIII) (Figs. 15E, H). The sperm sacs, which are usually located at XII-XIII½, vary in position from XI-XII¼ to XIII-XV. The male and female openings, the penis sheath, and the vaginal stalk are usually left of the nerve cord, but one or more may be positioned to the right of the nerve cord. In some individuals there is a tendency for the posterior crop caeca to be confluent with the intestine between XIX and XX, resulting in an apparently shortened intestine. The intestine usually begins at XIX, but the intestine of one individual began at XX. The ordinarily large, well-developed albumen gland is sometimes small and poorly developed, and the common oviduct varies from being straight to much coiled. The relatively small, compact epididymis, which is usually along the entire length of the sperm sac, often protrudes posteriorly beyond the end of the sperm sac (Figs. 15A, B) but is never so pronounced as in *H. grandis*. The condition of the epididymis, common oviduct, and albumen gland, and the position of the sperm sacs and male and female openings relative to the nerve cord, appear to be relatively unreliable systematic characters in this species.

Haemopis marmorata, a moderate-sized species represented in the present study by individuals from 2.0 to 8.0 cm, has a remarkably flaccid body that drapes limply between the fingers when picked up.

Ecology. In Ontario and Michigan *H. marmorata* is known to make mass movements or migrations upstream in late spring (Richardson, 1942; Sawyer, 1970b), but the significance of these movements is not yet clear. Other aspects of the behavior, neuroanatomy, and neurophysiology of this species were investigated by Miller in a series of papers from 1935 to 1945.

In the daytime this amphibious leech is usually found only partially submerged in water under large rocks and logs at the water's edge. In the laboratory also it will rest with part or all of its body out of the water and will often crawl completely out of its container. At night I have observed *H. marmorata* at the shoreline eating small invertebrates, especially pulmonates (*Physa*), slugs, and oligochaetes. Young individuals eat only the soft parts of the snails, leaving the shells, whereas larger individuals eat the entire snail, shell and all. Stomach contents included partially digested pulmonates, their shells, and large oligochaetes. Like most *Haemopis*, *H. marmorata* is a predator and scavenger rather than a bloodsucking parasite and has been reported eating oligochaetes, insect larvae, pelecypods, dead fish, and other leeches (*Helobdella stagnalis*, *Erpobdella punctata*, *Dina dubia*, and other *H. marmorata*) (Moore, 1912, 1924b; Moore, 1966). In spite of the reports that it will attach to man (Moore, 1912; Beck, 1954), there is no evidence that this species will actually suck the blood of humans.

Large *H. marmorata* were often encountered with from one to many individuals of both *Helobdella stagnalis* and *Glossiphonia complanata* attached to their backs. Judged from their engorged guts, the latter but not the former were feeding. A few individuals were found heavily infested with metacercariae as in earlier reports (Meyer and Moore, 1954).

The breeding habits of this species, which lays its eggs in sclerotized cocoons deposited in the mud or damp earth along the shores of lakes or streams (Mathers, 1948), are practically unknown but probably closely resemble those described in detail for *H. kingi* by Mathers (1954). Newly hatched young have been found from 30 June to mid-October (Moore, 1922; Moore, 1966), suggesting that breeding occurs from late spring to early summer.

Distribution (Fig. 34). *Haemopis marmorata* is the most widely distributed and most commonly encountered hirudinid in North America. Published records believed reliable include Alaska (Moore, 1898; Moore and Meyer, 1951), Northwest Territories (Moore, 1964), British Columbia (Clemens *et al.*, 1939; Scudder and Mann, 1968), Alberta (Bere, 1929; Moore, 1964; Clifford, 1969), Saskatchewan (Moore and Meyer, 1951; Oliver, 1958), Manitoba (Meyer and Moore, 1954), Ontario (Faull, 1913; Ryerson, 1915; Moore, 1924b, 1936; Rawson, 1930; Meyer and Moore, 1954), Quebec (Meyer and Moore, 1954), Newfoundland (Blanchard, 1896b; Pawlowski, 1948), Nova Scotia (Pawlowski, 1948), Prince Edward Island (Moore, 1922; Richardson, 1943), New Brunswick (Richardson, 1943), St. Pierre and Miquelon (Blanchard, 1896b), Wyoming (Moore, 1898), Utah (Verrill, 1874b; Beck, 1954), Colorado (Verrill, 1874b; Herrmann, 1970), New Mexico (Verrill, 1874b), Kansas (Moore, 1898), Nebraska (Ward, 1902), Iowa (Mathers, 1948), Minnesota (Moore, 1912), Wisconsin (Cahn, 1915; Bere, 1931; Sapkarev, 1968), Michigan (Leidy, 1868; Hankinson, 1908, 1916; Miller, 1937), Illinois (Moore, 1901), Missouri (Meyer, 1937a), Indiana (Moore, 1920), Ohio (Miller, 1929), New York (Moore, 1923), Connecticut (Verrill, 1874a), and Pennsylvania (Moore, 1912). The finding of this species in Delaware in the present study is the first record for that state.

Its occurrence west of Utah and Wyoming and south of Kansas, Missouri, and Delaware has not yet been documented.

Haemopis grandis (Verrill, 1874)

Semiscolex grandis Verrill, 1874a:672; Forbes, 1890a:119; Moore, 1952:3.
Haemopis grandis: Ruthven, 1906:51; Moore, 1912:116, figs. 26-28, 37; Ryerson, 1915:166; Moore, 1918:658; Moore, 1922:8; Moore, 1923:15, pl. 1; Moore, 1924b:29; Bere, 1929:177; Miller, 1929:10; Rawson, 1930:36; Bere, 1931:439; Meyer, 1937b:118; Miller, 1937:85; Townes, 1937:167; Richardson, 1943:90; Mathers, 1948:419; Moore, 1952:3; Pennak, 1953:317, fig. 201L; Mathers, 1954:460; Meyer and Moore, 1954:91; Rupp and Meyer, 1954:294; Oliver, 1958:163; Moore, 1959:555; Mann, 1961b:165; Mathers, 1963:168; Moore, 1964:2; Moore, 1966a:11; Thomas, 1966:202.
Mollibdella grandis: Richardson, 1969:117, figs. 3D, 6D.

Description (Fig. 13E). The uniform ground color of typical *H. grandis* varies from slate gray or light brown to a dark green and has from a few to many irregularly spaced black blotches. The dorsum is usually, but not always, somewhat darker and has more blotches than the venter, which may be pigment-free. There is some variation in the number of blotches from almost complete absence to 25 or even more. In at least one population the dorsal blotching approached the heavy mottling found in *H. marmorata* and led momentarily to misidentification with that species.

There are no signs of metameric pigment patterns in the young or adults. Especially in small individuals there may often be broad yellow marginal bands, but they are usually not as distinct as those in the more heavily pigmented *H. terrestris*, which usually has a conspicuous black middorsal stripe not found in *H. grandis*.

The penis, which is short (9-10 mm) and thick, is commonly protruded in preserved specimens. The gonopores are often in, or almost in, furrows (b5/b6) but occur just as frequently, even within the same population, on the anterior third or, more uncommonly, in the middle of the rings (b6) (Fig. 16B).

Dissections were made on seven individuals of *H. grandis* representing four localities in Michigan and ranging in size from 2.9 to 10.5 cm (Figs. 15F, I, Table 3) (Sawyer, 1968). Distinctive characteristics of this species are an unusually large, loose epididymis extending caudad well beyond the tip of the sperm sac (Fig. 15C), the position of the anterior part of the prostate gland (XI$\frac{1}{4}$-XII, usually XI$\frac{1}{2}$), the posterior flexion of the penis sheath (XII-XIV$\frac{1}{2}$, usually XIII), the ratio of the short and long arms of the penis sheath (1:1.0 to 1:1.8), the position of the center of the ovaries (XII$\frac{1}{2}$), the posterior extension of the vaginal system (XII$\frac{3}{4}$-XV, usually XIV$\frac{1}{2}$), and the posterior extension of the narrow posterior crop caeca (XXI$\frac{1}{4}$-XXIII$\frac{1}{4}$, usually XXII). The sperm sacs, which are usually located at XI$\frac{1}{2}$-XIII, vary in position from XI-XII$\frac{1}{2}$ to XII$\frac{1}{2}$-XIII$\frac{3}{4}$. The male and

TABLE 3

VARIATIONS OF REPRODUCTIVE AND DIGESTIVE SYSTEMS OF *H. GRANDIS*
(7 individuals from Michigan)

Length (cm)	8.4	10.5	7.8	6.7	6.7	4.0	2.9
Color	Heavily mottled	Heavily mottled	Heavily mottled	Typical (few blotches)	Typical	Typical	Typical
Posterior Extension:							
penis sheath	XIV	XIII	XIV½	XIII	XIII	XIII	XII
vaginal stalk	XIV½	XIV½	XV½	XIV½	XV	XIV¾	XII¾
crop caeca	XXI½	XXI¼	XXIII½	XXII	XXII1½	XXI½	XXIII¼
Ratio of Short and Long Arms of Penis Sheath	1:1.0	1:1.3	1:1.0	1:1.1	1:1.7	1:1.8	1:1.0
Position:							
center of ovary	XII½	XII½	XII½	XII½	XII½	XII½	XII½
anterior edge of prostate gland	XI½	XI½	XI¼	XI½	XII	XI½	XI½
sperm sacs, left	XI¾–XIII	XI¾–XII¾	XI¼–XIII½	XI–XIII	XII–XIV	XI–XII¾	—
sperm sacs, right	XI½–XII½	XII½–XIII½	XII½–XIII½	XII–XIII¾	XI¾–XIII½	XI–XIII	XI¼–XII
Relative to Nerve Cord:							
male opening	Right	Right	Left	Right	Right	Left	Right
female opening	Right	Right	Left	Left	Right	Right	Right
penis sheath	Right	Center	Left	Right	Right	Right	Right
vaginal stalk	Right	Center	Left	Left	Left	Right	Right

female openings, penis sheath, and vaginal stalks are usually to the right of the nerve cord, but one (usually the female system) or more can be positioned left of the nerve cord. The intestine usually begins at XIX, but in one individual it began slightly anterior at XVIII¾. The albumen gland is usually large and elongate, but the common oviduct varies from short and coiled to long and straight. The condition of the common oviduct and the albumen gland, and the position of the sperm sacs and the male and female openings relative to the nerve cord, are subject to so much variation that they are relatively unreliable systematic characters in this species.

Haemopis grandis, a large species represented in the present study by individuals from 2.9 to 10.5 cm long, has a flaccid body which drapes limply between the fingers when picked up.

Ecology. This jawless and toothless species, a predator and scavenger rather than a parasite, has never been known to suck human blood. Stomach contents included pulmonate and bivalve shells and other leeches (*Placobdella ornata*), and it has been reported eating oligochaetes, pulmonates, insect larvae, and other leeches (*Macrobdella decora*) (Moore, 1912, 1922, 1923; Ryerson, 1915; Rupp and Meyer, 1954). Its habits are similar to those of *H. marmorata*. Almost nothing is known about reproduction in this species.

Distribution (Fig. 35). *Haemopis grandis* appears to have a more northern and eastern distribution than *H. marmorata*. Published records believed reliable include Alberta (Bere, 1929), Saskatchewan (Oliver, 1958), Manitoba (Meyer and Moore, 1954), Ontario (Ryerson, 1915; Moore, 1924b; Rawson, 1930; Moore, 1936; Meyer and Moore, 1954; Thomas, 1966), Quebec (Meyer and Moore, 1954; Richardson, 1969), New Brunswick (Richardson, 1943), Prince Edward Island (Richardson, 1943), Minnesota (Moore, 1912), Wisconsin (Bere, 1931; Miller, 1937), Michigan (Miller, 1937), Lake Erie (Miller, 1929), New York (Moore, 1923), and Maine (Rupp and Meyer, 1954). Its occurrence in the United States west of Wisconsin and south of Wisconsin, Michigan, Ontario, and New York has not been documented. Unfortunately, some reports in the literature may actually be *H. plumbea* or *H. marmorata*, both of which can closely resemble *H. grandis* externally.

Haemopis terrestris (Forbes, 1890), new combination

?*Hirudo lateralis* Say, 1824:15 (name not assignable with certainty).
Semiscolex terrestris Forbes, 1890a:119; Forbes, 1890b:646.
Haemopis lateralis: Moore, 1898:560; ?Moore, 1901:528, figs. 25, 27-32; Ward, 1902:275; ?Moore, 1912:113, fig. 23; ?Andrews, 1915:200; Moore, 1918:658; Miller, 1929:10; Meyer, 1937a:250; ?Miller, 1937:85; ?Mathers,

1948:397, pl. 3; Pennak, 1953:317, fig. 201J; Mathers, 1954:460; Moore, 1959:555; Mann, 1961b:164, fig. 17B; Mathers, 1963:168.
Haemopis lateralis terrestris: Moore, 1918:649.
Haemopsis lateralis: ?Mullin, 1926a:61, pl. III, fig. 3.
Haemopis laterallis: Miller, 1937:88.
Percymoorensis lateralis: Richardson, 1969:121.

Description (Fig. 13F). Although the uniformly dark ground color, which completely lacks dark blotches or heavy mottling both dorsally and ventrally, is perhaps the most consistent aspect of the pigmentation of *H. terrestris*, its most characteristic aspect is a conspicuous middorsal black stripe from the eyes to the anal region and a yellowish stripe along the margins on either side from the neck to the anal region. The middorsal stripe varies in preserved specimens from a thin, slightly interrupted stripe to a relatively thick band, and in some populations this stripe may even be absent. The lateral stripes vary in preserved specimens from a barely perceptible light area along the margins to a strong, well-defined, intensely bright stripe. The gonopores, which usually occur on the rings, are separated by five to five and a half annuli, their actual relative positions being subject to some variation (Fig. 16A). The penis, commonly protruded in preserved specimens, is remarkably long (25-30 mm) and slender (Fig. 14F).

Dissections of six individuals of *H. terrestris* from four localities in Illinois (Table 4, Figs. 15G, J) (Sawyer, 1969) agreed well with the findings of Moore (1901, fig. 27). The relatively invariable characters of considerable systematic significance in this species include the position of the anterior part of the prostate gland (XII-XII$\frac{1}{4}$), the posterior flexion of the penis sheath (XIII$\frac{3}{4}$-XIV$\frac{1}{4}$, usually XIV), the ratio of the short and long arms of the penis sheath (1:1.2 to 1:1.6, usually 1:1.6), the center of the ovaries (XII-XII$\frac{1}{2}$), the posterior part of the vaginal system (XIII$\frac{3}{4}$-XV, usually XIV), and the extension of the large posterior crop caeca (XXIII-XXIV), the left and right caeca sometimes extending to different levels.

Other internal characters usually regarded as systematically important are variable and are less reliable in defining this and probably related species, including the condition of the relatively short, coiled common oviduct, the large elongate albumen gland, and the relatively loose epididymis. The sperm sacs, which are relatively small (Fig. 15D), are usually located at XI$\frac{1}{4}$-XII$\frac{1}{4}$ but vary in position from X$\frac{3}{4}$-XII to XII-XII$\frac{1}{2}$. The male gonopore and penis sheath are usually to the right of, and the female counterparts left of, the nerve cord, but either can lie to the right or left. One dissected individual lacked all of the female organs except for the ovaries.

TABLE 4

VARIATIONS OF REPRODUCTIVE AND DIGESTIVE SYSTEMS OF *H. TERRESTRIS*

(6 individuals from Illinois)

LENGTH (cm)	12.3	15.2	13.8	9.1	10.0	9.5
COLOR	No middorsal stripe	No middorsal stripe	No middorsal stripe	Typical	Typical	Typical
POSTERIOR EXTENSION:						
penis sheath	XIV	XIII¾	XIV	XIV¼	XIV	XIV¼
vaginal stalk	XIV	XIII¾	XIV	Absent	XV	XIV¼
crop caeca	XXIII	XXIV	l. XXIII r. XXIV	—	l. XXIII½ r. XXIII	XXIII
RATIO OF SHORT AND LONG ARMS OF PENIS SHEATH	1:1.6	1:1.5	1:1.6	1:1.2	1:1.6	1:1.4
POSITION:						
center of ovary	XII¼	XII	XII¼	XII¼	XII½	XII¼
anterior edge of prostate gland	XII¼	XII	XII¼	XII	XII	XII
sperm sacs, left	XI–XII¼	XI–XII	XI–XII¼	XI½–XII½	XI–XII	XI¼–XII¼
sperm sacs, right	X¾–XII	XI–XII	XI–XII¼	XII–XII½	XI¼–XII¼	XI–XII
RELATIVE TO NERVE CORD:						
male opening	Right	Right	Left	Right	Left	Right
female opening	Left	Left	Left	—	Left	Right
penis sheath	Right	Right	Left	Right	Left	Right
vaginal stalk	Left	Left	Center	Absent	Right	Left

Haemopis terrestris, a relatively large species represented in the present study by individuals from 8 to 15 cm long, has a firm body, correlated with its terrestrial burrowing habits.

Remarks. Say's vague and perhaps erroneous description of *Hirudo lateralis*, an obscure leech from Minnesota, has led to a number of nomenclatural difficulties (Moore, 1952). Verrill (1872) assigned the name to an erpobdellid, *Nephelis lateralis*, probably known today as *Erpobdella punctata*. Moore (1898), however, considered *Hirudo lateralis* to be a hirudinid identical with Forbes's (1890) terrestrial leech *Semiscolex terrestris*, a species that may not occur at Say's type-locality. In 1912 Moore speculated on the possibility that *Hirudo lateralis* was also a partial synonym of *Haemopis plumbea* Moore, 1912. The confusing description of *Hirudo lateralis*, which may fit a number of species (*Haemopis terrestris*, *H. plumbea*, *H. grandis*, *H. marmorata*, and even one or two erpobdellids), should be regarded as inadequate, and the use of the name *lateralis* should cease. The name *Haemopis terrestris*, then, becomes the earliest available name for Forbes's well-described terrestrial leech.

For some time *H. terrestris*, then known as *H. lateralis*, was synonymized with a similar Chilean terrestrial leech, *Americobdella valdiviana* (Philippi), but upon a close examination of the two species it was determined beyond doubt that they represented two distinct, unrelated species (Moore, 1898, 1924a; Caballero, 1941; Soós, 1966a).

Ecology. *Haemopis terrestris* is unique among American leeches in being truly terrestrial, occurring in damp soil, usually under rocks and logs and well away from the water. The existence and systematic standing of an aquatic variety reported by earlier authors (Moore, 1912; Miller, 1929) need to be investigated. It is not known whether *H. terrestris*, normally a predator and scavenger, is ever parasitic. Stomach contents in the present study consisted of large oligochaetes, an observation which agrees well with Forbes (1890). Almost nothing is known about its reproduction.

Distribution (Fig. 35). Because of possible confusion with related species, little reliance can be put on some earlier published reports, but those believed reliable include Ohio (Miller, 1929), Illinois (Forbes, 1890; Moore, 1898, 1901), southeastern Missouri (Meyer, 1937a), and northwestern Tennessee (Moore, 1898). There is no real evidence that *H. terrestris* occurs as far north as Minnesota, as was thought by Moore (1912), who synonymized it with the inadequately described aquatic species *Hirudo lateralis* Say, 1824, from Minnesota. It may be restricted mainly to the area between the Mississippi and Ohio rivers and bordering areas, with a center of distribution in Illinois and with northern limits in southern Wisconsin and Michigan, but a more extensive and critical examination of the systematics and distribution of this species is needed.

Haemopis plumbea Moore, 1912

Haemopis plumbeus Moore, 1912:115, figs. 29-31; Moore, 1918:658; Mullin, 1926a:62, pl. III, fig. 4; Miller, 1929:10; Miller, 1937:85; Mathers, 1948: 397, pl. 3; Pennak, 1953:317, fig. 201K; Mathers, 1954:460; Meyer, 1968:17.
?*Haemopsis plumbeus:* Mullin, 1926a:36, pl. III.
Haemopis plumbeous: Miller, 1937:87.
Haemopis plumeus: Mathers, 1948:412.
Haemopis plumbea: Moore, 1959:555; Mann, 1961b:165; Mathers, 1963:168.
Bdellarogatis plumbeus: Richardson, 1969:117, figs. 3E, 4A, 6C.

Remarks. *Haemopis plumbea*, a rare leech not encountered in the present study, is a little-known species of uncertain systematic standing, differing in internal anatomy (see key) from *H. grandis* and *H. marmorata*, which it can closely resemble externally. Not enough is yet known about its distribution and systematic standing to make generalizations; it is known from Minnesota (Moore, 1912), Iowa (Mullin, 1926; Mathers, 1948), Wisconsin (Miller, 1937), Michigan (Miller, 1937), Ohio (Miller, 1929), and Quebec (Richardson, 1969).

GENUS *MACROBDELLA* VERRILL, FEBRUARY 1872
(NOT *MACROBDELLA* PHILIPPI, OCTOBER 1872)

The well-known genus *Macrobdella* is unique among American hirudinids in having characteristic copulatory glands located on the ventral surface at XIII/XIV and XIVb1/b2, about ten and eleven annuli behind the male gonopores (Figs. 14B, C). Our understanding of this distinctive genus, which contains one of the first leeches to be described from North America, has undergone little modification since its description by Verrill (1872b). *Macrobdella* is represented by three moderately large (5-10 cm) species from North America. The northern *M. decora* (Say, 1824) is the only species known from the midwestern states. The southern *M. ditetra* Moore, 1953, may be found in the extreme southern tip of Illinois. The rare *M. sestertia* Whitman, 1886, is known only from Cambridge, Massachusetts.

Macrobdella decora (Say, 1824)

Hirudo decora Say, 1824:267; Leidy, 1868:229; Leidy, 1870:89.
Hirudo ornata: Ebard, 1857:55; Verrill, 1874a:688; Moore, 1952:4.
Macrobdella decora: Verrill, 1872b:138, fig. 4; Verrill, 1874a:668; Moore, 1898:561; Moore, 1901:508, figs. 22-23; Ward, 1902:274; Hankinson, 1908:

232; Moore, 1912:106, figs. 24-25, 38; Ryerson, 1915:166; Hankinson, 1916: 118; Moore, 1918:656; Moore, 1922:8; Moore, 1923:15, pl. 1C, figs. 12-14; Moore, 1924b:28; Mullin, 1926a:36, pl. III, fig. 2; Miller, 1929:10, fig. 1; Moore, 1936:114; Meyer, 1937b:118; Miller, 1937:85; Richardson, 1942: 68; Richardson, 1943:89; Mathers, 1948:397, pls. 3-4; Pawlowski, 1948:332, figs. 2-3; Caballero, 1952:203; Moore, 1952:3; Moore, 1953:8 (not p. 12); Pennak, 1953:317, fig. 201B; Mathers, 1954:466; Meyer and Moore, 1954: 91; Rupp and Meyer, 1954:294; Moore, 1959:553, fig. 23.8; Cargo, 1960: 119, fig. 1; Mann, 1961b:163, fig. 13; Gouck *et al.*, 1967:959; Sawyer, 1968:228; Richardson, 1969:105, figs. 1B, 5D; Herrmann, 1970:5.

Description (Fig. 13A). This moderate-sized (5-9 cm) hirudinid has a copulatory zone consisting of two rows of two copulatory gland pores each, located at XIII/XIV and XIVb1/b2, ten and eleven annuli behind the male gonopore (Fig. 14C). Its distinctive, brightly colored dorsum has about 20 metameric middorsal red dots and corresponding lateral black dots on a uniform dark green background. The ventral surface is reddish and usually has a few scattered black splotches. The posterior half of the ventral surface of the posterior sucker is usually heavily pigmented with black. The male and female gonopores, located at XI/XII and XII/XIII respectively, are usually separated by five annuli, but some populations were found in southeastern Michigan in which the male opening was situated slightly on the next posterior ring, XIIb1, its position in the closely related *M. sestertia*. The position of the female gonopore was invariable in the specimens examined, but Moore (**1912**) reported its occurrence on XIIIb1. Moore (**1922**) found an individual of *M. decora* from Algonquin Park, Canada, with only three copulatory glands, the left posterior one missing.

Ecology. Macrobdella decora, commonly called the American medicinal leech because of its extensive use in medicine for many years, is the notorious bloodsucking leech frequently encountered by swimmers in the northern United States and Canada. In some places it can be such a problem to swimmers that swimming must be restricted or even discontinued. Such a heavy infestation of *M. decora* in Palisades Interstate Park, New York, was the cause of an extensive study by Moore (**1923**) on its natural history to find a means of controlling outbreaks of this species.

This leech, which is rarely found in flowing water or large open lakes, often abounds in small temporary and permanent ponds, as well as in heavily vegetated, mud-bottomed marshes and ditches. Its extremely sharp teeth and sanguinivorous habits allow it to pierce the skins and suck the blood of a number of vertebrates, including man, cattle, turtles, frogs, toads (Moore, **1923**, fig. 13B; Brockleman, **1968**), fish (sturgeon: Moore, 1924b; trout: Rupp and Meyer, 1954), and wading birds

(Mathers, 1948). In addition, it is a voracious predator, feeding on eggs of various amphibians, oligochaetes (tubificids and earthworms), insect larvae, other *M. decora*, and snails (Moore, 1923; Mathers, 1948). Published analyses of stomach contents have revealed large numbers of tubificids, occasional insect larvae (Ward, 1902; Moore, 1912), and vestiges of salamander eggs (Cargo, 1960). Behavioral observations of its highly developed sensitivity to chemical and tactile stimulation, especially in relation to finding food, were made by Whitman (1886), Moore (1923), and Gouck *et al.* (1967). *Macrobdella decora* is known to engorge itself in spring and early summer on aggregations of such spawning vertebrates (especially on their eggs) as frogs (*Rana catesbeiana*), toads (*Bufo americanus*) (Moore, 1923, figs. 13B, C; Brockleman, 1968), salamanders (*Ambystoma maculatum*) (Cargo, 1960, fig. 1), and trout (*Salvelinus fontinalis*) (Rupp and Meyer, 1954). This predation in the spring may account for as much as 80 percent of the egg mortality in toads (*Bufo americanus*) and probably other species (Brockleman, 1968).

Engorgement in the spring appears to be a prerequisite to breeding, which has been described in part by Moore (1923) and Gouck *et al.* (1967). The straw-colored elliptical cocoons, made of a spongy chitinoid material characteristic of the family, are laid (in New York) in June or July in the mud under logs and rocks at the water's edge, the newly emerged young being encountered in July and August (Moore, 1923, fig. 14). Under laboratory conditions Gouck *et al.* (1967) found that cocoons were laid between one and two months after feeding, and that after 28-30 days an average of 16 young about 20-22 mm long emerged. Mathers (1948) reported only about eight young per cocoon.

Distribution (Fig. 36). *Macrobdella decora* appears to be primarily a northern species, especially abundant from Colorado to Saskatchewan, northward to the Georgian Bay, eastward to Maine and the Maritime Provinces, and southward to Kansas, Illinois, Virginia, and Maryland. Published records believed reliable include Saskatchewan (Moore, 1922; Meyer and Moore, 1954), Ontario (Ryerson, 1915; Moore, 1922, 1924b, 1936; Meyer, 1937b), Quebec (Moore, 1922; Richardson, 1969), Nova Scotia (Moore, 1922), New Brunswick and Prince Edward Island (Richardson, 1943), Colorado (Herrmann, 1970), Kansas (Verrill, 1874a), Nebraska (Ward, 1902), Iowa (Mathers, 1948), Minnesota (Moore, 1912), Wisconsin (Miller, 1937), Michigan (Adams, 1908; Hankinson, 1908, 1916; Miller, 1937; Sawyer, 1968), Illinois (Moore, 1901), Ohio (Miller, 1929), Pennsylvania (Rathbun, 1884; Moore, 1901, 1912, 1923), extreme western Virginia (Moore,

1898), Maryland (Cargo, 1960), New York (Moore, 1898, 1923; Miller, 1929; Barrow, 1953), Connecticut (Verrill, 1874a; Barrow, 1953), and Maine (Verrill, 1874a; Rupp and Meyer, 1954).

In the Illinois Natural History Survey collection is a vial containing an *M. decora* labeled "eight miles northwest of Monte Morelos, Nuevo Leon, Mexico," which corroborates the existence of an intriguing, apparently disjunct population of *M. decora* in Nuevo León in northern Mexico, first reported by Caballero (1952).

Macrobdella ditetra Moore, 1953
Macrobdella ditetra Moore, 1953:5, pl. 1, fig. 1; Brandt, 1936:502; Moore, 1959:553, fig. 23.8; Mann, 1961b:163, fig. 13; Sawyer, 1967:32; Meyer, 1968:18.
Macrobdella decora: Moore, 1953:12.

Description (Fig. 13B). The specific name *ditetra*, which was used by ecologists (Brandt, 1936) long before the species was described by Moore in 1953, is based on a unique characteristic of this species, a copulatory zone with two rows of four copulatory gland pores each, located at XIII/XIV and XIVb1/b2, ten and eleven annuli behind the male gonopore (Fig. 14B). The gland pores of mature individuals are well developed, but those of immature individuals are small and can easily be overlooked. Typical individuals are dark green on the dorsum with a paramarginal and a supramarginal row of dark dots on each side. Unlike *M. decora*, there is no middorsal row of metameric red dots. The dots of the internal or supramarginal row are usually fused into a dark longitudinal stripe, but in some individuals they may be poorly developed or almost nonexistent. The dots of the paramarginal row are irregularly spaced and are usually less developed than the supramarginal row, but the vestiges of the former can be found on almost all individuals, even when the supramarginal row is missing. In some heavily pigmented individuals there is also a wide dark longitudinal band extending middorsally from the region of the eyes to the anal region, but usually this band is very faint. The venter is a uniform cream color and usually, but not always, has irregular dark splotches concentrated near the margins.

Ecology. During a study of frog parasites in eastern North Carolina, Brandt (1936) observed that *M. ditetra* commonly infested bullfrogs (*Rana catesbeiana*) only in midsummer and showed a strong preference for large bullfrogs over 100 mm long. In the early spring *M. ditetra* feeds on frog eggs, which apparently stimulate breeding activity (Moore, 1953). It has never been reported to attack humans, but judged from its sanguinivorous congenitor *M. decora*, it should be expected to do so. Little else is known about feeding or reproduction in this species.

Meyer (1959) reported that during routine milking of a dairy cow in Florida, two moderate-sized (60 × 5 mm) individuals of *M. ditetra* were found in the teats.

Distribution. *Macrobdella ditetra* is a southern coastal-plain species previously reported from Texas, Louisiana (Sawyer, 1967), Alabama, South Carolina (Moore, 1953), North Carolina (Brandt, 1936), and Florida (Meyer, 1959). In the present study a vial containing a single individual from McIntosh County, Georgia, was found, which is the first record for that state. It has not yet been found in the midwestern United States.

Macrobdella sestertia Whitman, 1886

Macrobdella sestertia Whitman, 1886:378, figs. 57-59; Moore, 1918:656; Moore, 1923:17; Pennak, 1953:317; Moore, 1959:553, fig. 23.8; Mann, 1961b:163, fig. 13.
Macrobdella testertia: Moore, 1953:7.

Remarks. A vial from the Harvard collection labeled "MCZ 1729, Chebaco row, 20 July 1875" and presumably from around Cambridge, Massachusetts, contained a single faded specimen of *M. sestertia* (Fig. 37). The copulatory glands were poorly preserved but did fit the original description, as did the gonopore separation of two and a half annuli. The dorsum was faded except for a faint paramarginal row of metameric black dots. Having finally seen a specimen of this rare species, which has remained unknown since its original description, I feel confident that *M. sestertia* does represent a recognizable morphological type, the systematic standing of which needs to be investigated.

GENUS *PHILOBDELLA* VERRILL, 1874

Verrill established *Philobdella* as a subgenus of *Macrobdella*, primarily on the basis of the remarkable external genital region, characterized by glandular adhesive organs containing gland pores, and copulatory depressions around the gonopores (Fig. 14A) (Moore, 1959, fig. 23.9; Mann, 1961b, fig. 16). By 1898 Moore had elevated *Philobdella* to full generic rank, which is undoubtedly its true systematic position. The type-species *P. floridana* Verrill, 1874, from Lake Okeechobee, Florida, is known only from the original description. Moore (1898) reported a species of *Philobdella* from Louisiana which he took at the time to be *P. floridana*, but later (1901) in a detailed morphological description he recognized it as a new species, *P. gracilis*.

Although *P. floridana* has not been reported since the original description almost a century ago, most authors continue to distinguish *P. floridana* and *P. gracilis* as separate species, primarily on the basis of pigmentation and number of teeth (see key). There is considerable doubt, however, that two or only one species of *Philobdella* is represented in the southern states.

Philobdella floridana Verrill, 1874

Macrobdella (Philobdella) floridana Verrill, 1874a:669; Moore, 1952:3.
Philobdella floridana: Moore, 1901:518; Moore, 1918:657; Moore, 1952:3; Pennak, 1953:317; Moore, 1959:554; Mann, 1961b:165; Sawyer, 1967:33.

Remarks. Philobdella floridana, not encountered in the present study, remains unknown since its original description from Lake Okeechobee, Florida.

Philobdella gracilis Moore, 1901

Philobdella floridana: Moore, 1898:561; Sawyer, 1967:33.
Philobdella gracile Moore, 1901:511, figs. 12-21; Moore, 1918:657; Pennak, 1953:317, figs. 201F, G; Viosca, 1962:243; Meyer, 1968:19.
Philobdella gracilis: Moore, 1952:5; Moore, 1953:4; Moore, 1959:554; Mann, 1961b:165, fig. 16; Sawyer, 1967:33.

Description (Fig. 13C). This moderately large (6 cm) bloodsucking hirudinid has a conspicuous middorsal light yellow stripe and a dorsal paramarginal row of irregularly spaced black dots on either side. Between the middorsal stripe and each row of dots are two brownish-black longitudinal bands, a thick one adjacent to the stripe and a narrow one nearer the row of black dots. The two bands are confluent just anterior to the anal region. The dorsum is basically dark except for the middorsal yellow stripe, which may be poorly developed in some individuals. The venter is lighter and has irregular dark splotches concentrated near the margins. The unique external genitalia of adult members of this genus and species are characterized by a copulatory pit or depression around each gonopore and a prominence or adhesive organ containing several conspicuous gland pores immediately anterior to each gonopore (Fig. 14A). The morphology and function of the external genitalia were described by Verrill (1874a) for *P. floridana* and by Moore (1898, 1901) for *P. gracilis*, both of whom reported that the adhesive organs and copulatory depressions apparently secure the two individuals together during mating.

Ecology. Although Viosca (1962) reported the result of an accumulation of observations on *P. gracilis* extending over a ten-year period, little is known about the biology of this species. He reported finding it attached to, but not necessarily feeding on, the following animals: frogs (*Rana catesbeiana, R. grylio, R. clamitans,* and *R. pipiens*), alligator (*Alligator mississippiensis*), snakes (*Agkistrodon piscivorus, Natrix cyclopion,* and *N. fasciatus*), and turtles (*Chelydra serpentina* and *Kinosternon subrubrum hippocrepis*). That *Philobdella* also feeds on earthworms was suggested by Verrill (1874), who reported an individual preserved in the process of eating a lumbricoid worm, and was corroborated by Moore (1901), who found *Allolobophora* in the gut. Like its near relatives *Macrobdella decora* and *M. ditetra, P. gracilis* is known to feed voraciously on frog eggs. It attacks especially those of *Rana pipiens*, one of the first to lay its eggs in the spring. In view of its well-developed jaws and teeth, it is rather surprising that *P. gracilis* has never been known to attack human beings.

Distribution (Fig. 36). *Philobdella gracilis*, represented in the present study by two vials from Illinois, one of which was not labeled, is a southern species known from Louisiana (Moore, 1898; Viosca, 1962) and southern Illinois (Moore, 1901), with the closely allied form *P. floridana* from southern Florida (Verrill, 1874a).

GENUS *HIRUDO* LINNAEUS, 1758

Hirudo medicinalis Linnaeus, 1758

Remarks. The European medicinal leech *Hirudo medicinalis*, which is now practically extinct in parts of Europe, was at one time imported for medicinal purposes into the northeastern United States by the thousands and even artificially cultured for a while (Hessel, 1881, 1884). During the early part of this century there was speculation that this species had escaped and established itself in the northeastern states, but it now seems more likely that the medicinal leech had been confused with the common horse leeches (*Haemopis* spp.). To my knowledge there has never been in this century a confirmed record of a wild population of *H. medicinalis* in the United States or Canada. Unless the occurrence of this species is confirmed, it is best to consider it as not established in North America.

ZOOGEOGRAPHICAL AND EVOLUTIONARY CONSIDERATIONS

The only certain fossil leeches are two species described from the upper Jurassic of Bavaria (Kozur, 1970). Except for the somewhat dubious *Pontobdellopsis cometa* described by Ruedemann (1901) from Albany, New York, there are no known fossils from North America. Our understanding of the evolutionary history of North American leeches — the manner and rapidity of speciation and dispersal, relative success in numbers and kinds before, during, and after the glacial periods, and historical reasons for modern distributions — must, therefore, be inferred from such indirect sources of evidence as host-parasite relationships and geographical distributions of extant species. Of the 19 genera represented in North America, including Mexico, eight are endemic: *Actinobdella, Illinobdella, Oligobdella, Piscicolaria, Nephelopsis, Mooreobdella, Macrobdella,* and *Philobdella*.

The occurrence of the following intricate host-parasite relationships suggests that at least some are of long evolutionary standing: among *Batracobdella picta*, trypanosomes, and amphibians (Richardson, 1949; Barrow, 1953; Brockleman, 1968, 1969; Woo, 1969); *Placobdella pediculata* and the drum (*Aplodinotus grunniens*) (Hemingway, 1912); *Theromyzon* and various species of birds (Sooter, 1937; Meyer and Moore, 1954; Moore, 1964, 1966a); *Piscicola punctata* and various teleosts (Thompson, 1927; Richardson, 1948); and *Piscicola salmo-*

sitica, hemoflagellates, and salmonid fish (Becker, 1964; Becker and Katz, 1965, 1966).

Many of the known species of North American leeches fall naturally into four groups: a group of widely distributed, almost ubiquitous species, a group of predominantly northern species, another of predominantly southern species, and a group of geographically restricted populations and species. The first group of ubiquitous species, e.g., *Helobdella stagnalis, Erpobdella punctata, Glossiphonia complanata,* and *G. heteroclita,* are among the most common leeches in North America. They are exceedingly adaptable and easily dispersed, and they have such wide ecological tolerances and catholic feeding habits that they shed little light on the problems of leech zoogeography.

The second group of northern species, e.g., *Oculobdella lucida, Placobdella ornata, Nephelopsis obscura, Dina dubia, D. parva, Mooreobdella fervida, Haemopis grandis, H. marmorata,* and *Macrobdella decora,* appear to be physiologically restricted to waters just above freezing point for extended periods of time. The southern limit of their range is roughly that of the glacial drift border of the Quaternary ice advances (Wright and Frey, 1965). The great expanses of northern North America, covered by ice until rather recently, were undoubtedly colonized for the most part by these leeches after the Wisconsinan ice advance from 10,000 to 70,000 years ago (Wright and Frey, 1965: 359). Of special interest are Prince Edward Island in eastern Canada and Kodiak Island, Alaska, both of which were covered by the Wisconsinan ice. Since that time these islands, which are today isolated from the mainland by the sea, have been colonized by at least eight (Richardson, 1943) and two (Moore and Meyer, 1951) species respectively. Similarly, Sable Island, approximately 150 miles east of Halifax, Nova Scotia, has three species of leeches (Gates and Moore, 1970). The colonization of the northern expanses by the relatively few northern species, which abound in the cold lakes and streams from Alaska to eastern Canada and as far north as Great Slave Lake (Moore and Meyer, 1951) and the Georgian Bay (Ryerson, 1915), is a case of animal dispersal of considerable magnitude.

The third group of southern species, e.g., *Placobdella multilineata, Helobdella lineata, Mooreobdella microstoma, Philobdella gracilis,* and *Macrobdella ditetra,* abound in the warm waters of the southeastern states. *Philobdella gracilis,* a characteristic southern species, extends only as far north as the southern tip of Illinois, the southernmost limit of the glacial drift border, but others extend as far north as southern Wisconsin and Michigan or even farther.

The fourth group of geographically restricted populations and species may suggest clues to the process of speciation of leeches in North America. The systematically obscure forms *Mooreobdella bucera, Macrobdella sestertia,* and *Philobdella floridana,* which are distinct morphological types on the periphery of the ranges of the more widely distributed and better-known species *Mooreobdella fervida, Macrobdella decora,* and *Philobdella gracilis* respectively, probably represent incipient species or populations adapted to marginal levels of existence (Mayr, 1965). If the occurrence of an apparently disjunct population of *Macrobdella decora* in northern Mexico (Caballero, 1952) represents a relict population, then the range of *M. decora,* an otherwise northern species, was at one time much larger than it is today. The biannulate species *Oligobdella biannulata,* which occurs on salamanders in the southern Appalachians (Moore, 1900; Sawyer, 1971b), may have its nearest affinities in and around Japan and in New Zealand. This species, which promises to shed light on the early evolution of the Rhynchobdellae, needs to be investigated.

The manner and likelihood of leech dispersals must vary considerably with the species, depending upon its size, behavior, hosts, ecology, and physiological requirements. Some species, *Haemopis marmorata* and *Erpobdella punctata,* are known to make mass movements upstream (Richardson, 1942; Sawyer, 1970), but most cases of dispersal seem to depend upon animal hosts. A variety of hosts capable of being useful dispersal agents, e.g., fish, turtles, mammals, amphibians, and especially birds, have been reported carrying *Theromyzon rude* (Meyer and Moore, 1954; Moore, 1964, 1966b), *T. meyeri* (Sooter, 1937), *Placobdella ornata* (Moore, 1964), *Helobdella stagnalis* (Moore, 1924b), and *Haemopis* sp. (Mullin, 1926), but the actual roles they have played in past dispersals are obscure. The interesting cases of hirudiniasis, such as the occurrence of *Nephelopsis obscura* in the air bladder of a lake trout (Meyer and Bangham, 1950), *Batracobdella picta* in the dorsal subcutaneous lymph spaces of a bullfrog (Richardson, 1949), and *Macrobdella ditetra* in the teats of a cow (Meyer, 1959), suggest the various and often bizarre ways that leeches may be dispersed. The available evidence suggests that it is the adult or juvenile, rather than the egg stage, that is probably involved in most cases of dispersal. There is little evidence that man has significantly altered the distribution of any North American species.

Considering the relatively large number of endemic genera and species, their wide distributions, and their often intricate host-parasite relationships, it seems likely that at least some of the North American leeches constitute a group of relatively long evolutionary standing.

KEY

Earlier keys and aids to identification of the freshwater leeches of all or parts of the United States and Canada can be found in Verrill (1874a), Moore (1912, 1918, 1959), Miller (1929, 1937), Meyer (1940, 1946a), Mathers (1948), Pennak (1953), Mann (1961b), Moore (1964, 1966a), and Hoffman (1967), but most of these are already outdated and incomplete. It is hoped that the following key, which combines as many characters as possible, will serve as a practical guide to identification of known freshwater species, especially for biologists unfamiliar with leeches. For convenience the family Piscicolidae, which is not discussed in the text, is included in the key.

For the most part, external and biological characters are sufficient for the identification of most American species, but identification of *Haemopis* and most erpobdellids, especially *Mooreobdella* and *Dina*, requires dissection. The most important external characters for identifying leeches are the number and arrangement of eyes, the presence and arrangement of jaws, papillae, sensillae, ocelli, and pulsatile vesicles, the pigmentation patterns, the size and general shape of the body, and the number of annuli per segment and between gonopores. Useful biological characters include hosts, swimming capability, the manner of moving and caring for eggs and young, and ecological and geographical variations.

To prevent severe muscular contractions during preservation, the leeches should first be relaxed by slowly adding 70 percent ethanol to the container until all movement stops. After the mucus is removed with a paper tissue, the leeches are then placed in a dissecting tray, the larger individuals being pinned in the narcotized position and covered with the fixative, usually formalin, to prevent softening of the tissue. After the tissue is hard, usually from 30 minutes to several hours or rarely longer, depending upon the size of the specimens, they are placed in the final preservative, usually 70 percent ethanol. If fixed properly, most of the pigments remain indefinitely, but the green pigment dissolves quickly in ethanol. The eyes are best examined by pressing a glass slide on the head region. The annuli are best examined in living material because preservation tends to distort secondary and tertiary subdivisions.

Whole mounts (flattened between two glass slides, either stained with borax carmine or dehydrated directly in a graded series of alcohols beginning with 70 percent, cleared in xylene, and then mounted in Canada balsam) as well as complete transverse and longitudinal series (cut at 10 μ and stained with Ehrlich's haematoxylin and eosin) are useful for morphological studies of smaller leeches. Larger erpobdellids and hirudinids should be dissected by pinning them at each end to the bottom of a dissecting dish before submerging them in 70 percent ethanol. Two dorsolateral longitudinal cuts should be made through the body wall, after which the cuts are joined by a transverse incision, allowing the dorsal portion of the body wall to be lifted off. The muscles and botryoidal tissue are removed with a fine forceps until the digestive and reproductive systems as well as the ganglionic guidelines are clearly visible. In the case of hirudinids a midventral slit beginning at the buccal cavity will reveal the jaws, which may be hidden in crypts of tissue.

In addition to the internal anatomical features which are drawn and labeled in Figs. 2, 5, 11, and 15, the following terms and abbreviations are used in the key.

Annuli are body rings, usually demarcated by metameric pigment patterns (Fig. 14). The conventional formula 7(14) means that there is a faint subdivision of each of seven annuli. The mid-body segments of most leeches have three primary annuli, labeled by convention as a1, a2, and a3 (or sometimes written a1-3). Each of these in turn can be further subdivided into the secondary annuli, b1, b2, b3 . . . b6 (or b1-6), and still further into the tertiary annuli, c1, c2, c3 . . . c12 (or c1-12). The neural annulus refers to the annulus in which the ganglion is located, usually a2 (or b3 + 4).

Ganglion is a concentration of nerve cell bodies in the ventral nerve chain. Excluding the subesophageal mass (six ganglia) and the caudal mass (seven ganglia), there are 21 ganglia in the ventral chain, labeled in Roman numerals, VII-XXVII. The neural annulus of segment 12 would, therefore, be expressed as XIIa2. The abbreviation VIII/IX refers to the somite limits between segments 13 and 14 (Fig. 15).

Gonopore is the external opening of either the male or female reproductive system, located on the midventral line about one-third the distance from the head (Figs. 9, 11, 14, and 16). Usually the female gonopore is located one complete segment posterior to the more prominent male gonopore (in the hirudinids this usually means five annuli posterior to the male gonopore).

Key to Species

1. Free-living or attached to turtles, amphibians, birds, or fish; body often excessively flattened; young often brooded by parent; blood colorless; 2, 3, or 6 (or more) annuli per complete segment; eyes 1-4 pairs; mouth a small pore on oral sucker; protrusible proboscis; no jaws or teeth; move by placing hind sucker immediately behind oral sucker in "inchworm" fashion; rarely if ever swim (RHYNCHOBDELLA)......................2
— Usually free-living; body elongate, not depressed, large (2½-20 cm); young never attached to ventral surface of parent; blood red; usually 5 annuli per complete segment; eyes 3-5 pairs; mouth capacious, occupying most of oral sucker cavity; jaws and teeth present or absent; no proboscis; may or may not move in "inchworm" fashion; good swimmers.......3
2. Rarely free-living, usually found attached to fish (the only nonpiscicolids commonly found on live fish are *Placobdella pediculata* on the drum and *Actinobdella triannulata* on the sucker); slender and elongate, often with lateral vesicles; body often more or less divided into narrow anterior and wider posterior regions; young never attached to ventral surface of parent; cocoons attached to substrate, never brooded; 6 or more (rarely 3) annuli per complete segment; oral sucker distinct from neck; eyes 0-3 pairs (Piscicolidae) (see Meyer, 1940, 1946a; Hoffman, 1967).......................................27
— Commonly free-living or attached to turtles, amphibians, or birds; body flattened, never cylindrical or with lateral vesicles; young always attached to ventral surface of parent;

eggs in membranous sacs, either attached to ventral surface of parent or to substrate and covered by parent's body; almost always 3 (rarely 2 or 6) annuli per complete segment; oral sucker ventral and more or less confluent with neck; eyes 1-4 pairs, usually 1 pair (Glossiphoniidae).................4
3. Eyes 3-4 (never 5) pairs, in 2 transverse rows (Fig. 17F), never in 1 parabolic arch; predacious, rarely if ever parasitic; body solid and worm-like, moderate size (2-5 cm); swim readily when disturbed; usually encountered completely submerged in water, not with part of body half out of water under large objects at water's edge; gonopores separated by 2-3½ annuli (Fig. 9); usually 5 annuli per complete segment, but may be further subdivided, especially in *Nephelopsis;* no true penis or external copulatory glands; no true jaws or teeth; long, weakly muscularized pharynx; no caeca from crop; testes small, arranged in bundles; never move by placing hind sucker behind oral sucker in "inchworm" fashion (PHARYNGOBDELLA, Erpobdellidae).............................32
— Eyes always 5 pairs, forming a parabolic arch (Fig. 17G); usually predacious (*Macrobdella* and *Philobdella* are bloodsuckers); body large (3-12 cm), soft, usually becoming limp and inactive when disturbed (except *Haemopis terrestris*); usually encountered with body half out of water under large objects at water's edge; gonopores usually separated by 5-5½ annuli (2-4 in some *Macrobdella* and *Philobdella*) (Fig. 14); always 5 annuli per complete segment; protrusible filiform penis or conspicuous copulatory glands around (or 10-11 annuli posterior to) the gonopores (Fig. 14); usually jaws with teeth; short muscular pharynx; always a pair of posterior crop caeca; testes large, segmentally arranged, usually 10 pairs; can but does not always move in "inchworm" fashion (GNATHOBDELLA, Hirudinidae)............................38
4. Eyes 3 pairs (Figs. 17A, B) (*Glossiphonia*)...................5
— Eyes 4 pairs (Fig. 17D) (the bird leeches, *Theromyzon*).....6
— Eyes 1 pair (Figs. 17C, E)................................7
5. Eyes 3 pairs, equidistant, in 2 longitudinal rows (Fig. 17A); a pair of narrow dark paramedial stripes dorsally and ventrally; a pair of dorsal metameric white dots paramedially and marginally; body opaque, internal organs not visible through integument; very common (Fig. 1A)......................
.....................*Glossiphonia complanata* (Linnaeus, 1758)

— First pair of eyes always closer than two posterior pairs (Fig. 17B); essentially no pigmentation dorsally or ventrally; body translucent, internal organs visible through integument; uncommon (Fig. 1B)...*Glossiphonia heteroclita* (Linnaeus, 1761)
6. Eyes 4 pairs, equidistant, in 2 longitudinal rows (Fig. 17D); body gelatinous, translucent (globular and opaque in recently fed individuals), with many fine black chromatophores dorsally and ventrally; 2 paramedial pairs and usually a marginal pair of bright yellow dots; often encountered on birds; 2 forms of obscure systematic standing:
— Gonopores separated by 2 annuli; apparently distributed in central and eastern United States (Fig. 1C).............
........................*Theromyzon meyeri* (Livanow, 1902)
— Gonopores separated by 3 annuli; apparently distributed in central and western United States and Canada............
.............................*Theromyzon rude* (Baird, 1863)
7. Large, almost hemispherical posterior sucker separated from body by a narrow pedicel (Fig. 1D); a circle of 30-60 retractile digitate processes with accessory adhesive gland ducts projecting into sucker cavity a short distance from its inner margin (Fig. 1E); a single pair of large eyes either touching or very close together; 1-3 series of dorsal papillae; mid-body segments 3-6 annulate; diffused salivary glands; small (about 1 cm); rare and poorly known (*Actinobdella*)................10
— Posterior sucker not unusually large or on a narrow pedicel (except *Placobdella pediculata* (Fig. 3B)); no circle of retractile processes projecting into the sucker cavity..............8
8. Eyes well separated (Fig. 17E); no metameric pigment patterns along lateral margins; 6 (or 1) pairs of gastric caeca; egg sacs always carried on ventral surface, never attached to substrate; small (1 cm); parasitic on snails; common, usually free-living (Figs. 4, 5).......................................21
— Eyes close together or touching (Figs. 2, 17C); usually small metameric whitish patches along lateral margins and a large whitish area around eyes; 7 pairs of gastric caeca; egg sacs on ventral surface or attached to substrate; small to large (1.0-6.5 cm)..9
9. Eyes close together but usually not touching (Fig. 17C); no conspicuous white ring in neck region or white patches in genital and anal regions; mouth on anterior rim of oral sucker; egg sacs attached to substrate, never carried on ventral sur-

face; moderate to large (2.0-6.5 cm); usually encountered on turtles (except *Placobdella pediculata,* on fish) or free-living (Fig. 3) (*Placobdella*)...................................12
— Eyes usually touching; conspicuous white ring in neck region and usually white patches in genital and anal regions; mouth within oral cavity and not on anterior rim of sucker; egg sacs attached to ventral surface of parent; small to moderate (1.0-2.5 cm); parasitic on amphibians; often free-living..18
10. Six unequal annuli per complete segment; eyes united; body slender, strongly convex......................................11
— Three equal annuli per complete segment; eyes close together, not usually united; body broader and flatter than alternatives; 29-31 moderate-sized conical papillae along inner margin of caudal sucker; known only from Ontario; free-living and from common sucker (*Catostomus commersoni*)...
.........................*Actinobdella triannulata* Moore, 1924
11. With 60 very small papillae along inner margin of caudal sucker; known only from Long Point, Lake Erie, from original description; from snapping turtle (*Chelydra serpentina*).....
..........................*Actinobdella annectens* Moore, 1906
— With 29-30 long pointed finger-like papillae along inner margin of caudal sucker; known from Illinois, Minnesota, and Ohio; free-living, host unknown (Figs. 1D, E)..............
......................*Actinobdella inequiannulata* Moore, 1901
12. Either 3 dorsal keel-like ridges and a wide discoidal head set off from rest of body by a narrow neck constriction (Fig. 3A), or dorsum smooth and caudal sucker set off from rest of body by a narrow peduncle (Fig. 3B); both uncommon............13
— Neither of above combinations of characters; usually on turtles or free-living; common.............................14
13. A wide discoidal head set off from rest of body by a narrow constriction; 3 keel-like ridges on dorsal surface, composed of uniform large pointed tubercles on every annulus; few if any tubercles or papillae between ridges; widely distributed but uncommon (Fig. 3A).........*Placobdella montifera* Moore, 1906
— Almost always encountered on drumfish (*Aplodinotus grunniens*); caudal sucker set off from rest of body by a long narrow peduncle (absent in juveniles smaller than 1 cm); dorsum smooth, nonpapillated; anus anteriorly positioned at XXIII/XXIV rather than usual XXVII/XXVIII; rare, known only from midwestern United States (Fig. 3B)..................
......................*Placobdella pediculata* Hemingway, 1908

14. Dorsum, in particular the middorsal line, with few if any papillae; dorsal pigmentation characterized by irregular intricate patterns 15
    — Dorsum usually heavily papillated, in particular the middorsal line; dorsal pigmentation unpatterned or characterized by longitudinal stripes and bands, especially middorsally 16
15. Venter with 8-12 bluish-green longitudinal stripes, without fine black chromatophores; no accessory eyes; often green; dorsum with a middorsal cream-colored stripe or band of variable width and with irregular lateral patches; very large (4-6 cm); never swims as adult; usually on turtles but often free-living, especially in July and August; widely distributed and common (Fig. 3C) .... *Placobdella parasitica* (Say, 1824)
    — Venter with numerous small dark chromatophores and without 8-12 bluish-green longitudinal stripes; 2 pairs of variable concentrations of dark pigment, always situated 2 and 5 complete annuli behind the single functional pair of eyes, giving the false impression of 3 pairs of eyes (accessory eyes) (Fig. 17C); rarely green; dorsum generally checkered with an interrupted broad middorsal and a smaller paramedial reddish-brown band; body ribbon-shaped; often swims as adult; usually free-living; uncommon (Fig. 3D) ....
    .... *Placobdella hollensis* (Whitman, 1892)
16. Dorsum with 5-7 distinct longitudinal rows of large metameric pointed papillae; other papillae small and inconspicuous; hind sucker with single circular row of papillae; light-colored longitudinal stripes on either side of middorsal row of papillae, joining in neck region; narrow continuous (or slightly interrupted) dark stripe encompassing middorsal row of papillae; venter with a pair of bluish longitudinal stripes and without fine dark chromatophores; body opaque (Fig. 3F) .... *Placobdella papillifera* (Verrill, 1872)
    — Dorsum entirely covered with numerous large rounded papillae which are not usually metameric or in 5-7 longitudinal rows; each annulus with 16-20 papillae of varying size; hind sucker usually without row of papillae; broad brown middorsal band; ventral surface with fine dark chromatophores and no pair of bluish longitudinal stripes; body translucent; two closely related northern and southern forms................. 17
17. From northern United States and Canada; middorsal band usually interrupted; dorsum warty; papillae large, numerous,

and irregularly positioned; ventral chromatophores scattered (Fig. 3E)....................*Placobdella ornata* (Verrill, 1872)
— From southern United States; middorsal band usually continuous; dorsum less rough; papillae less numerous, smaller, with the larger ones tending to be in 5 indistinct longitudinal rows; ventral chromatophores tending to be in many longitudinal rows.................*Placobdella multilineata* Moore, 1953

18. Body segments biannulate; known from mountains of North Carolina and South Carolina; parasitic on salamander *Desmognathus*................*Oligobdella biannulata* (Moore, 1900)
— Body segments triannulate (Fig. 2) (*Batracobdella*).........19

19. No white patches in genital and anal regions and at most only slight, but variable, white patches around eyes and in neck region; body smooth, opaque; no middorsal or marginal metameric dots, prominences, or papillae; indistinct dark middorsal stripe; two paramedial pairs of yellowish metameric dots; only glossiphoniid commonly encountered on mating frogs, toads, and salamanders; usually in small mud- and leaf-bottomed ponds (Fig. 2C)...............*Batracobdella picta* (Verrill, 1872)
— White patches in eye, neck, genital, and anal regions; body translucent; conspicuous middorsal and marginal metameric dots or papillae..20

20. Body convex, thick; in addition to marginal dots, metameric pigmentation consisting only of 3 series of usually dark-tipped papillae; sometimes with a slightly flattened body, a thick dark band encompassing middorsal series of papillae, and a white patch approximately halfway between genital and anal patches; uncommon, usually along shores of large lakes and rivers (Figs. 2D, E)..........*Batracobdella phalera* (Graf, 1899)
— Body excessively flattened, thin; no true dark-tipped papillae; 5 longitudinal rows of white prominences surrounded by yellowish dots equidistant longitudinally and transversely; known only from southern Michigan (Figs. 2A, B)..........
............................*Batracobdella michiganensis* n. sp.

21. Dorsal and ventral surfaces heavily pigmented with uniform grayish-blue chromatophores and with thin dark paramedial lines extending to neck region; body opaque; dorsal surface smooth, no papillae or scute; gonopores united; anteriorly situated proboscis pore; uncommon, but locally abundant (Fig. 5E)............*Oculobdella lucida* Meyer and Moore, 1954
— Dorsum unpigmented or pigmented with longitudinal or

transverse stripes and metameric dots; papillated or with a chitinous scute in neck region; gonopores separated by at least 1 annulus; centrally located proboscis pore (*Helobdella*).......22
22. A chitinous scute or plaque in neck region (Fig. 17E); very common (Fig. 5C).........*Helobdella stagnalis* (Linnaeus, 1758)
— No such scute in neck region.............................23
23. Dorsum smooth, no papillae...............................24
— Dorsum with 3-7 longitudinal series of papillae on neural annulus; degree of papillation variable......................26
24. Body unpigmented, elongate, and cylindrical; lateral margins of body almost parallel; translucent, internal organs visible through body integument; 1 pair of crop caeca (Fig. 5D)....
...........................*Helobdella elongata* (Castle, 1900)
— Body pigmented with longitudinal or transverse stripes; flattened and leaf-shaped; opaque, internal organs not visible through body integument; 6 pairs of crop caeca...............25
25. Dorsum with transverse rusty-brown bands alternating with white bands, the latter consisting of 8-10 confluent white dots on neural annulus; no longitudinal pattern; pigment fades in ethanol; known only from present study from southwestern Michigan (Figs. 5A, B).............*Helobdella transversa* n. sp.
— Dorsum with 6 major longitudinal white stripes alternating with 6 coffee-brown stripes, including a middorsal band; no transverse pattern; pigment remaining after preservation in ethanol; uncommon, from large lakes and cold waters of northern United States and Canada (Figs. 4D-F)..............
..............................*Helobdella fusca* (Castle, 1900)
26. Dorsum with 3 series of small black-tipped papillae; 4 series of metameric white dots on neural annulus external to papillae; no middorsal dots; a variable species, sometimes with longitudinal stripes or with reduced number of papillae; common, especially in southern states and in warm water (Figs. 4B, C)
..............................*Helobdella lineata* (Verrill, 1874)
— Dorsum roughly papillated, with many whitish rounded papillae arranged in 5-7 longitudinal series on each neural annulus; dorsum whitish, usually unpigmented; relatively uncommon (Fig. 4A)........*Helobdella papillata* (Moore, 1906)
27. Without pulsatile vesicles along lateral margins of body region; 1 pair of eyes on oral sucker; caudal sucker usually smaller than body width; body not separated into distinct neck (trachelosome) and body regions; 5 pairs of testes........28

— With 11 pairs of pulsatile vesicles (not very conspicuous in *Piscicola*) along lateral margins of urosome; usually 2 pairs of eyes on oral sucker; caudal sucker as wide as or wider than body width; body may or may not be separated into distinct neck and body regions; 6 pairs of testes......................29

28. A conspicuous pair of narrow black paramedial stripes extending from eyes to anal region; body stout, flattened centrally, and convex dorsally; small oral and caudal suckers, the latter distinct from body; mid-body segments triannulate; uncommon (Fig. 18D)................*Piscicolaria reducta* Meyer, 1940
 — No longitudinal stripes; body elongate, narrow, and cylindrical; small oral and caudal suckers, the latter more or less confluent with body and posteriorly directed; mid-body segments 12 (14) annulate; common (several forms known, possibly representing one variable species; see Meyer, 1940, 1946) (Fig. 18C)....................*Illinobdella moorei* Meyer, 1940

29. Pulsatile vesicles large and conspicuous, even after preservation; body sharply separated into 2 regions, a small narrow trachelosome and a wide urosome; caudal sucker large, oral sucker relatively small; mid-body segments 7 annulate (Fig. 18A).......................*Cystobranchus verrilli* Meyer, 1940
 — Pulsatile vesicles small and obscure, may be overlooked after preservation; body elongate, not sharply separated into neck and body regions; caudal sucker moderately large, about twice width of oral sucker; mid-body segments 12 (14) annulate, may be 6 (7) in juveniles (*Piscicola*)....................30

30. No ocelli on caudal sucker; gonopores separated by 4 tertiary annuli; known east of Rocky Mountains (Fig. 18B)........
 ....*Piscicola punctata* (Verrill, 1871)
 — With 8-12 ocelli on caudal sucker; gonopores separated by 2 annuli...31

31. With 8-10 ocelli on caudal sucker; sperm duct much convoluted; known from western United States and Canada......
 ....*Piscicola salmositica* Meyer, 1946
 — With 10-12 (and perhaps more) ocelli on caudal sucker; sperm duct simply looped; known east of Rocky Mountains
 ....*Piscicola milneri* (Verrill, 1871)
 (= ?*Piscicola virginica* (Hoffman, 1964), new combination; = ?*Piscicola geometra* of Moore (1898), Bere (1931), and Mason *et al.* (1970). The occurrence of *P. geometra* in North America is questionable.)

32. Ejaculatory duct with a pre-atrial loop extending anteriorly to ganglion XI (Fig. 9) 35
— Ejaculatory duct without a pre-atrial loop extending anteriorly to ganglion XI (Fig. 9) (*Mooreobdella*) 33
33. Gonopores separated by 3 annuli, usually in furrows (Fig. 9C); atrial horns laterally projecting; color usually uniform smoky gray to brownish, without black pigment; distributed south of southern tip of Great Lakes; common (Fig. 8E) ....
.... *Mooreobdella microstoma* (Moore, 1901)
— Gonopores separated by 2 annuli; atrial horns more anteriorly projecting than laterally; distributed in northern United States and Canada 34
34. Atrial horns anteriorly projecting (Fig. 9E); gonopores usually on rings; color uniform smoky gray, sometimes with minute black chromatophores; moderate size (2-4 cm); distributed in northern tier of states and Canada; common in cold-water streams and lakes (Fig. 8G) ....
.... *Mooreobdella fervida* (Verrill, 1874)
— Atrial horns more laterally projecting than anteriorly (Fig. 9D); gonopores on rings or in furrows; color uniform smoky gray without black pigment; nerve commissure appears through integument as a white ring around neck; slightly smaller than alternative (2-3 cm); known only from southeastern Michigan; uncommon, but locally abundant in small semipermanent wood ponds (Fig. 8F) ....
.... *Mooreobdella bucera* (Moore, 1949)
35. Gonopores separated by 2 annuli, usually in furrows; all midbody annuli of equal size and not subdivided, or all annuli of varying width and subdivided 1 or more times 36
— Gonopores separated by 3½ (2½-4) annuli, usually on rings; every fifth annulus in mid-body region wider and slightly subdivided (*Dina*) 37
36. A paramedial and sometimes a paramarginal pair of variable black pigment concentrations, forming 2 or 4 black longitudinal stripes; mid-body annuli not subdivided; male gonopore in adults remarkably large (Fig. 9F); very common throughout most of United States and Canada (Figs. 8C, D) ....
.... *Erpobdella punctata* (Leidy, 1870)
— Dorsum greenish brown, covered with sparse scattered black blotches; no longitudinal stripes; most mid-body annuli partially subdivided once or twice; male gonopore in adults smaller than alternative (Fig. 9G); distributed only in north-

ern tier of states and Canada; fairly common in cold-water lakes and streams (Fig. 8H).... *Nephelopsis obscura* Verrill, 1872
37. Gonopores separated by 3½ or more (3½-4) annuli (Fig. 9A); heavily mottled dorsum, often with a variable black middorsal stripe; greenish color quickly dissolves in ethanol; moderate size (2-6 cm); uncommon, but locally abundant (Fig. 8A).................. *Dina dubia* Moore and Meyer, 1951
— Gonopores separated by 3½ or fewer (2½-3½) annuli (Fig. 9B); body virtually unpigmented; color uniform smoky gray; small, usually less than 2 cm, but occasionally much larger; uncommon (Fig. 8B)............ *Dina parva* Moore, 1912 (*Dina anoculata* Moore, 1898, a poorly known species described from California, keys out here, but it differs from *D. parva* in being without eyes.)
38. External copulatory glands, located 10-11 annuli posterior to male gonopore or located around gonopores, which are obscured by deep copulatory depressions; eversible male bursa but no true penis or penis sheath; gonopores usually separated by 2-4 annuli (5-5½ in *Macrobdella decora*) (Fig. 14)........39
— No external copulatory glands; gonopores never obscured by copulatory glands or depressions; gonopores separated by 5 annuli; filiform penis and penis sheath (Figs. 14D-F); long cylindrical pharynx with thin walls; adults with straight simple crop with no lateral caeca except for single pair of posterior crop caeca; common in northern United States and Canada, unknown from southern states (Figs. 13D-F) (*Haemopis*; see Richardson, 1969).............................43
— Six reddish longitudinal stripes; 3-5 cm in length; mammalian bloodsuckers; probably not now established in North America.................................... *Hirudo medicinalis*
39. Glandular area around gonopores; gonopores separated by 3-4 annuli, obscured by deep copulatory depressions (Fig. 14A); double row of teeth (distichodont) per jaw; long cylindrical pharynx with thin walls; straight simple crop with no lateral caeca except for single pair of posterior crop caeca (Fig. 13C) (*Philobdella*)...42
— External copulatory glands located about 10-11 annuli posterior to male gonopore (Figs. 14B, C); single row of teeth (monostichodont) per jaw; short bulbous muscular pharynx; lateral caeca in each segment of crop (Figs. 13A, B) (*Macrobdella*)...40
40. No median dorsal series of 20 metameric dots; 8 copulatory

glands (2 rows of 4); gonopores separated by 2 annuli (Fig. 14B); 46-55 teeth per jaw; distributed in southeastern United States; common (Fig. 13B)......*Macrobdella ditetra* Moore, 1953
— A median dorsal series of about 20 red metameric dots (Fig. 13A) 41

41. Four copulatory glands (2 rows of 2); gonopores usually separated by 5-5½ annuli (Fig. 14C); 50-65 teeth per jaw; distributed in northern United States and Canada, also known from northern Mexico (Caballero, 1952); very common (Fig. 13A).........................*Macrobdella decora* (Say, 1824)
— Twenty-four copulatory glands (2 rows of 2 groups, containing 6 glands each); gonopores separated by 2½ annuli; 39-46 teeth per jaw; heretofore known only from Cambridge, Massachusetts, from original description; rare (Fig. 37)......
.........................*Macrobdella sestertia* Whitman, 1886

42. Middorsal stripe, if present, dark brown; dorsum dark brown with two faint reddish-brown bands along each side toward margins, separated by a narrow black stripe; supramarginal band sometimes broken but no distinct spots; margins and venter dull reddish brown; about 20 teeth on each jaw; rare and known only from original description from southern tip of Florida..................*Philobdella floridana* (Verrill, 1874)
— Conspicuous light yellow middorsal stripe and a lateral row of irregularly spaced black dots; venter uniform light yellow except for some irregular dark splotches, especially near margins; about 40 (35-48) distichodont teeth on each jaw; common and widely distributed throughout southern states, extending up Mississippi Valley to southern tip of Illinois (Fig. 13C)....................*Philobdella gracilis* Moore, 1901

43. Annuli VIIa3 and VIIIa1 completely subdivided ventrally, i.e., 27 distinct annuli from oral cavity to annulus (XIb6) bearing male gonopore (Fig. 14F); middorsal black stripe and yellowish marginal stripes; body firm; terrestrial to semi-aquatic; common, known from Illinois, Ohio, southeastern Michigan, and probably neighboring states; large, 8-15 cm (4-19 cm); jaws with teeth; flexion of penis sheath at XIV (XIII¾-XIV¼);* anterior edge of prostate gland at XII-XII¼ (Figs. 15G, J); ratio of short and long arms of penis

---

* In the key to the various species of *Haemopis,* positions are pinpointed by fractional distances between known ganglia; i.e., X¾ refers to three-quarters of the distance between ganglia X and XI, or halfway between the somite limit X/XI and ganglion XI.

sheath 1:1.6 (1:1.2 to 1:1.6) (Fig. 13F).................
........................*Haemopis terrestris* (Forbes, 1890)
—Annuli VIIa3 and VIIIa1 not subdivided ventrally, i.e., only 25 distinct annuli from oral cavity to annulus (XIb6) bearing male gonopore (Figs. 14D, E); no middorsal black stripe (except *Haemopis kingi*); body soft and limp (except *H. kingi*); semiaquatic to aquatic; both rare and common forms; size variable, usually moderate (2-10 cm)............44

44. Jaws with teeth present; posterior crop caeca large, extending to XXIII-XXIV; color variable; ovaries at either XII-XII$\frac{1}{2}$ or XIII$\frac{3}{4}$ (XIII-XIV$\frac{1}{2}$); vaginal system extending to either XIV$\frac{1}{2}$ or XVI (XV-XVII) (Figs. 15E, H)..................45
—No jaws or teeth; posterior crop caeca thin, extending in adult to XXII (XXI$\frac{1}{4}$-XXII$\frac{1}{2}$) but perhaps farther (XXIII-$\frac{1}{4}$) in juveniles; color usually uniform slate gray with a few irregular black blotches, but some may be heavily mottled dorsally; often with yellowish marginal stripes; ovaries at XII-XII$\frac{1}{2}$; vaginal system extending to XIV$\frac{1}{2}$ (XII$\frac{3}{4}$-XV) (Figs. 15F, I)..47

45. Flexion of penis sheath at XVII (XVI-XVIII$\frac{1}{4}$); anterior edge of prostate gland at XIV (XIII$\frac{1}{4}$-XV$\frac{1}{2}$); vaginal system extending to XVI (XV-XVII); ovaries at XIII$\frac{3}{4}$ (XIII-XIV$\frac{1}{2}$) (Figs. 15E, H); very common, known from most of Great Lakes states, other northern states, and Canada; 2 color forms: (1) olive green with heavy mottling dorsally and ventrally (Fig. 13D); (2) uniform slate gray with a few irregular black blotches, resembling *H. grandis*....................
........................*Haemopis marmorata* (Say, 1824)
—Flexion of penis sheath at XII-XIII$\frac{1}{2}$; anterior edge of prostate gland at X$\frac{3}{4}$-XI; vaginal system extending to XIV-XIV$\frac{1}{2}$; ovaries at XII$\frac{1}{4}$-XII$\frac{1}{2}$............................46

46. Color olive green with moderate to heavy black blotching dorsally and with yellowish marginal stripes; no middorsal black stripe; body soft and limp; epididymis massive and extending well beyond posterior end of sperm sac; flexion of penis sheath at XIII$\frac{1}{4}$; size, 5-10 cm; known only from northwestern Iowa and southwestern Minnesota....................
....................*Haemopis lateromaculata* Mathers, 1963
—Color olive green with a middorsal black stripe and with yellowish marginal stripes; a few irregular black blotches; young with metameric black transverse bands; body firm; epididymis not especially massive or extending conspicuously

beyond sperm sac; flexion of penis sheath at XII¼; size, 6-13 cm..........................*Haemopis kingi* Mathers, 1954
47. Flexion of penis sheath at XIII (XII-XIV½) (Figs. 15F, I); anterior edge of prostate gland at XI½ (XI¼-XII); ratio of short and long arms of penis sheath 1:1.0 to 1:1.8; epididymis massive and extending well beyond posterior end of sperm sac (Fig. 15C); size, 3-11 cm (3-18 cm); common, widely distributed over most of northern United States and Canada (Fig. 13E)....................*Haemopis grandis* (Verrill, 1874)
— Flexion of penis sheath at XVI; anterior edge of prostate gland at about XIII¾; ratio of short and long arms of penis sheath 1:2.0; epididymis apparently not massive or extending conspicuously beyond sperm sac; size, 4-13 cm; uncommon, poorly known from original description from northern Minnesota and from a few references from other Great Lakes states and Canada...............*Haemopis plumbea* Moore, 1912

REFERENCES

A thorough attempt has been made to bring together the relevant primary literature which has contributed to our understanding of the biology of the leeches of North America (including the Piscicolidae), but some important works may have been overlooked. Since most of our knowledge of the distribution of leeches has come from a number of ecological and faunistic papers, the most important of these have also been included. The sources not read by the author are indicated by a single asterisk, and those which appear in the text but are not directly relevant to American species are indicated by a double asterisk. For leeches reported from North American freshwater fish, the reader is referred to the thorough treatment by Hoffman (1967).

Adams, Charles C. 1908. An ecological survey of Isle Royale, Lake Superior. A report from the University of Michigan Museum, published by the State Biological Survey, as a part of the report of the Board of the Geological Survey for 1908. Leeches, pp. 250-252.

Amin, Omar M. 1969. Helminth fauna of suckers (Catostomidae) of the Gila River system, Arizona. II. Five parasites from *Catostomus* spp. *American Midland Naturalist* 82(2):429-443.

Andrews, Olive V. 1915. An ecological survey of Lake Butte des Morts Bog, Oshkosh, Wisconsin. *Wisconsin Natural History Society Bulletin*, n.s. 13(4): 196-201.

Autrum, H. 1936. Hirudineen. In H. G. Bronns, *Klassen und Ordnungen des Tierreichs*. Vol. 4, sect. 3, bk. 4, pt. 1. Leipzig.

Baird, William. 1869. Descriptions of some new suctorial annelides in the collection of the British Museum. *Zoological Society of London Proceedings* 1869:310-318.

Baker, F. C. 1922. The molluscan fauna of the Big Vermilion River, Illinois, with special reference to its modification as a result of pollution by sewage and manufacturing wastes. Illinois Biological Monographs, vol. 7, no. 2. Urbana: University of Illinois. Hirudinea, p. 16.

———. 1924. The fauna of the Lake Winnebago region. *Wisconsin Academy of Sciences, Arts and Letters Transactions* 21:109-146.

Bangham, Ralph V. 1928. Parasites of black bass. *Scientific Monthly* 27:267-270.

———. 1933. Parasites of the spotted bass, *Micropterus pseudalplites* Hubbs, and summary of parasites of smallmouth and largemouth black bass from Ohio streams. *American Fisheries Society Transactions* 63:220-228.*

———. 1941. Parasites of fish of Algonquin Park lakes. *American Fisheries Society Transactions* 70:161-171.*

———. 1951. Parasites of fish in the upper Snake River drainage and in Yellowstone Lake, Wyoming. Scientific Contributions of the New York Zoological Society, *Zoologica* 36(3):213-217.*

———. 1955. Studies on fish parasites of Lake Huron and Manitoulin Island. *American Midland Naturalist* 53(1):184-194.*

Bangham, Ralph V., and J. R. Adams. 1954. A survey of the parasites of freshwater fishes from the mainland of British Columbia. *Fisheries Research Board of Canada Journal* 11(6):673-708.*

Bangham, Ralph V., and Carl E. Venard. 1942. Studies on parasites of Reelfoot Lake fish. IV. Distribution studies and checklist of parasites. *Tennessee Academy of Science Journal* 27(1):22-38.*

———. 1946. Parasites of fish of Algonquin Park lakes. *University of Toronto Studies in Biology* 65:33-46.*

Barrow, J. H. 1953. The biology of *Trypanosoma diemyctyli* (Tobey). I. *Trypanosoma diemyctyli* in the leech *Batrachobdella picta* (Verrill). *American Microscopical Society Transactions* 72:197-216.

Beck, D. E. 1954. Ecological and distributional notes on some Utah Hirudinea. *Utah Academy of Sciences, Arts and Letters Proceedings* 31:73-78.

Becker, Clarence D. 1964. The parasite-vector-host relationship of the hemoflagellate, *Cryptobia salmositica* Katz, the leech, *Piscicola salmositica* Meyer, and certain freshwater teleosts. Ph.D. thesis, University of Washington.

Becker, Clarence D., and Max Katz. 1965a. Transmission of the hemoflagellate, *Cryptobia salmositica* Katz, 1951, by a rhynchobdellid vector. *Journal of Parasitology* 51:95-99.

———. 1965b. Distribution, ecology and biology of the salmonid leech, *Piscicola salmositica* (Rhynchobdellae:Piscicolidae). *Fisheries Research Board of Canada Journal* 22(5):1175-95.

———. 1966. Host relationships of *Cryptobia salmositica* (Protozoa:Mastigophora) in a Washington hatchery stream. *American Fisheries Society Transactions* 95(2):196-202.

Becker, D. A., R. G. Heard, and P. D. Holmes. 1966. A preimpoundment survey of the helminth and copepod parasites of *Micropterus* spp. of Beaver Reservoir in north-west Arkansas. *American Fisheries Society Transactions* 95(1):23-34.

Bennike, S. A. B. 1943. Contributions to the ecology and biology of the Danish freshwater leeches. *Folia Limnologica Scandinavica* 2:1-109.

Bere, Ruby. 1929. Reports of the Jasper Park lakes investigations, 1925-26. III. The leeches. *Contributions to Canadian Biology and Fisheries* (Toronto), n.s. 4(14):177-183.

———. 1931. Leeches from the lakes of northeastern Wisconsin. *Wisconsin Academy of Science Transactions* 26:437-440.

Blanchard, Raphael. 1896a. XXI. Hirudinées. Viaggio del dot. A. Borelli nella republica Argentina e nel Paraguay. *Musei di Zoologia e di Anatomia Comparata della R. Università di Torino Bollettino*, vol. 11, no. 263.

———. 1896b. Courtes notices sur les Hirudinées. XXIII. Hirudinées de Terre Neuve et des iles adjacentes. *Société Zoologique de France Bulletin* (Paris), no. 21.*

———. 1896c. Courtes notices sur les Hirudinées. XXIV. Presence de la *Glossiphonia complanata* en Amerique. *Société Zoologique de France Bulletin* (Paris), no. 21.*

Bosc, Louis Augustine Guillaume. 1802. *Histoire naturelle des vers*. 3 vols. Paris. Sangsue, 1:232-251.*

Brandt, B. B. 1936. Parasites of certain North Carolina Salientia. *Ecological Monographs* 6:491-532.

Branson, Branley A., and Bobby G. Amos. 1961. The leech *Placobdella pediculata* Hemingway parasitizing *Aplodinotus grunniens* in Oklahoma. *Southwestern Naturalist* 6:53.

Bristol, Charles Lawrence. 1896. The classification of *Nephelis* in the United States. *New York Academy of Sciences* 3:33.

———. 1897. The metamerism of *Nephelis*. A contribution to the morphology of the nervous system, together with a description of *Nephelis lateralis*. *Zoological Bulletin* 1(1):35-39.

———. 1898. The metamerism of *Nephelis*. A contribution to the morphology of the nervous system with a description of *Nephelis lateralis*. *Journal of Morphology* 15:17-72.

Brockleman, Warren Y. 1968. Natural regulation of density in tadpoles of *Bufo americanus*. Ph.D. thesis, University of Michigan.

———. 1969. An analysis of density effects and predation in *Bufo americanus* tadpoles. *Ecology* 50(4):632-644.

Brooks, V. K. 1882. *Handbook of invertebrate zoology*. Boston.*

Caballero y C., Eduardo. 1931a. *Glossiphonia socimulcensis* n. sp. *Instituto de Biología Anales* (Mexico) 2:85-90.*

———. 1931b. Batrachobdellinae subfam. nv. *Instituto de Biología Anales* (Mexico) 2:223-229.**

———. 1940. Sobre la presencia de la *Placobdella rugosa* (Hirudinea:Glossiphonidae) en las aguas del Lago de Xochimilco. *Instituto de Biología Anales* (Mexico) 13(1/2).

———. 1941. Hirudineos de México. XVI. Nuevos huespedes y localidades para algunas sanguijuelas ya conocidas y descripcion de una nueva especie. *Instituto de Biología Anales* (Mexico) 12(2):752-753.*

———. 1952. Sanguijuelas de México. XVIII. Presencia de *Macrobdella decora* (Say, 1824) Verrill, 1872, en el Norte del Pais, y nueva desinencia para los ordenes de Hirudinea. *Instituto de Biología Anales* (Mexico) 23:203-209.

———. 1955. Hirudineos de México. XIX. Presencia de *Pontobdella macrothela* Schmarda, 1861, en aguas marinas del Golfo de Mexico. *La Escuela Nacional de Ciencias Biologicas Anales* (Mexico) 8(3/4):153-158. (English summary.)

———. 1956. Hirudineos de México. XX. Taxa y nomenclatura de la clase Hirudinea hasta generos. *Instituto de Biología Anales* (Mexico) 27:279-302.

———. 1959. Hirudineos de México. XXII. Taxa y nomenclatura de la clase Hirudinea hasta generos (nueva edición). *Instituto de Biología Anales* (Mexico) 30:227-242.

Cahn, A. R. 1915. An ecological survey of the Wingra Springs region, near Madison, Wisconsin, with special reference to its ornithology. *Wisconsin Natural History Society Bulletin* 13(3):123-177.

Cargo, D. G. 1960. Predation of eggs of the spotted salamander *Ambystoma maculatum* by the leech *Macrobdella decora*. Contribution 155, Maryland Department of Research and Education, Solomons, Maryland; *Chesapeake Science* 1(2):119-120.

Carlson, Clarence A. 1968. Summer bottom fauna of the Mississippi River, above Dam 19, Keokuk, Iowa. *Ecology* 49(1):162-169.

Castle, W. E. 1900a. Some North American fresh-water Rhynchobdellidae, and their parasites. Harvard College, *Museum of Comparative Zoology Bulletin* 36(2):17-64.

———. 1900b. The metamerism of the Hirudinea. *American Academy of Arts and Sciences Proceedings* 35:285-303.*

Chernin, Eli, et al. 1956. Studies on the biological control of schistosome-bearing snails. II. The control of *Australorbis glabratus* populations by the leech, *Helobdella fusca*, under laboratory conditions. *American Journal of Tropical Medicine and Hygiene* 5:308-314.

Clemens, W. A., et al. 1939. A biological survey of Okanagan Lake, British Columbia. *Fisheries Research Board of Canada Bulletin*, no. 61. Hirudinea, pp. 60, 68.*

Clifford, Hugh F. 1969. Limnological features of a northern brown-water stream, with special reference to the life histories of the aquatic insects. *American Midland Naturalist* 82(2):578-597.

Cope, O. B. 1958. Incidence of external parasites on cut-throat trout in Yellowstone Lake. *Utah Academy of Sciences, Arts and Letters Proceedings* 35:95-100.

Cordero, E. H. 1937. Hirudineos neotropicales y subantarcticos nuevos, criticos o ya conocidos del Museo Argentino de Ciencias Naturales. *El Museo Argentino de Ciencias Naturales Anales* (Buenos Aires), no. 39.

Cronk, M. W. 1932. The bottom fauna of Shakespeare Island Lake, Ontario. *University of Toronto Studies in Biology* 36:31-65. Leeches, p. 54.*

DeRoth, G. C. 1953. Some parasites from Maine freshwater fishes. *American Microscopical Society Transactions* 72(1):49-50.*

Diesing, C. M. 1850. Systema Helminthum. *Vindobonae* 1:435-471.

Earp, B. J., and R. L. Schwab. 1954. An infestation of leeches on salmon fry and eggs. *Progressive Fish-Culturist* 16:122-124.

Ebard, E. 1857. *Nouvelle monographie des Sangsues médicinales.* Paris.*

Eddy, S., and A. C. Hodson. 1945. *Taxonomic keys to the common animals of Minnesota exclusive of the parasitic worms, insects and birds.* Minneapolis: Burgess Publishing Co. Leeches, pp. 28-30.

Ernst, Carl H. 1971. Seasonal incidence of leech infestation on the painted turtle, *Chrysemys picta. Journal of Parasitology* 57(1):32.

Evermann, B. W., and H. W. Clark. 1920. Lake Maxinkuckee. Indiana Department of Conservation 1(7):304-305.

Faull, J. H. 1913. *The natural history of the Toronto region.* Washington, D.C.: Carnegie Institution. Hirudinea, p. 277.*

Forbes, S. A. 1890a. An American terrestrial leech. *Illinois State Laboratory of Natural History Bulletin* 3:119-122.

———. 1890b. An American terrestrial leech. *American Naturalist* 24:646-649.*

———. 1893. A preliminary report on the aquatic invertebrate fauna of the Yellowstone National Park, Wyoming, and of the Flat-Head region of Montana. *U.S. Fisheries Commission Bulletin* 1891:207-258. Hirudinea, pp. 218-220.

Fredeen, F. J. H., and J. A. Shemanchuk. 1960. Black flies (Diptera:Simuliidae) of irrigation systems in Saskatchewan and Alberta. *Canadian Journal of Zoology* 38:723-735.

Gates, G. E., and J. E. Moore. 1970. The freshwater and terrestrial Annelida. Appendix II (p. 45). Fauna of Sable Island and its zoogeographic affinities. National Museum of Natural Sciences (Ottawa), *Publications in Zoology* 4:1-45.

Gee, Wilson. 1913. The behaviour of leeches with especial reference to its modifiability. *University of California Publications in Zoology* 11:197-305.

Girard, Charles. 1850. *Phyllobranchus Ravenellii* Girard. *American Association for the Advancement of Science Proceedings* 4:124-125.*

Goldstein, R. J., and H. W. Wells. 1966. Note on the incidence of a marine leech, *Branchellion ravenellii* (*Gymnura micrura*, new host record). *Journal of Parasitology* 52(4):690.

Gouck, H. K., et al. 1967. Screening of repellents and rearing methods for water leeches. *Journal of Economic Entomology* 60(4):959-961.

Graf, Arnold. 1899. Hirudineenstudien. Deutschen Akademie der Naturforscher (Halle), *Nova Acta Abhandlung der Kaiserlich Leop.-Carol.* 72(2):215-404.

Grube, A., ed. 1850. *Die Familien der Anneliden, mit Angabe ihrer Gattungen und Arten. Archiv für Naturgeschichte* (Berlin), vol. 16. Discophora, pp. 354-364. (Also published separately, Berlin, 1851.)

———. 1871. Beschreibungen einiger Egelarten. *Archiv für Naturgeschichte* (Berlin) 37(1):87-121.

Gruffydd, L. D. 1965. Notes on a population of the leech, *Glossiphonia heteroclita,* infesting *Lymnaea pereger. Annals and Magazine of Natural History* 8(87/88):151-154.

Haderlie, E. C. 1953. Parasites of the freshwater fishes of northern California. *University of California Publications in Zoology* 57:303-440.*

Hagadorn, Irvine R. 1958. Neurosecretion and the brain of the rhynchobdellid leech, *Theromyzon rude* (Baird, 1869). *Journal of Morphology* 102:55-90.

———. 1961. Neurosecretory activity in the brain of a leech, *Theromyzon rude*. *American Zoologist* 1:451-452.

———. 1962a. Functional correlates of the neurosecretion in the rhynchobdellid leech, *Theromyzon rude*. *General and Comparative Endocrinology* 2:516-540.

———. 1962b. Neurosecretory phenomena in the leech, *Theromyzon rude*. *Society for Endocrinology Memoirs* 12:313-322 (H. Heller and R. B. Clark, eds., *Neurosecretion* (New York: Academic Press)).

———. 1966a. The histochemistry of the neurosecretory system in *Hirudo medicinalis*. *General and Comparative Endocrinology* 6(2):288-294.

———. 1966b. Neurosecretion in the Hirudinea and its possible role in reproduction. *American Zoologist* 6:251-261.

Hagadorn, Irvine R., et al. 1963. The fine structure of the supraesophageal ganglion of the rhynchobdellid leech, *Theromyzon rude*, with special reference to neurosecretion. *Zeitschrift für Zellforschung* 58:714-758.

Hagadorn, Irvine R., and R. S. Nishioka. 1961. Neurosecretion and granules in neurones of the brain of the leech. *Nature* 191(4792):1013-14.

Hale, John G. 1966. Influence of beaver on some trout streams along the Minnesota north shore of Lake Superior. *Minnesota Fisheries Investigation* 4:5-29.

Hall, F. G. 1922. The vital limits of exsiccation of certain animals. *Biological Bulletin* 42:31-51.

Hankinson, T. L. 1908. A biological survey of Walnut Lake, Michigan. A report of the Biological Survey of the State of Michigan, published by the State Board of Geological Survey as part of the report for 1907. Hirudinea, p. 232.

———. 1916. Shiras expeditions to Whitefish Point. Miscellaneous papers of the zoology of Michigan. In A. G. Ruthven, ed., *Michigan Geological and Biological Survey*. Publication 20, ser. 4. Hirudinea, pp. 118, 126, 132-137.

Harding, W. A. 1910. A revision of the British leeches. *Parasitology* 3(2):130-201.

Harms, Clarence E. 1960. Some parasites of catfishes from Kansas. *Journal of Parasitology* 36:695-701.

Hatto, Jane. 1968. Observations on the biology of *Glossiphonia heteroclita* (L.). *Hydrobiologia* 31:363-384.

Heldt, Thomas J. 1961. Allergy to leeches. *Henry Ford Hospital Medical Bulletin* 9(4):498-519.

Hemingway, Ernest E. 1904. The brain and nerve cord of *Placobdella pediculata*. *Science* 19:219.*

———. 1908. *Placobdella pediculata* n. sp. *American Naturalist* 42:527-532.*

———. 1912. The leeches of Minnesota. Part II. The anatomy of *Placobdella pediculata*. In Geological and Natural History Survey of Minnesota, *Zoology* 5:29-63.

Herrmann, Scott J. 1970. Systematics, distribution and ecology of Colorado Hirudinea. *American Midland Naturalist* 83(1):1-37.

Hessel, Rudolf. 1881. Artificial culture of medicinal leeches and of species of *Helix. U.S. Fisheries Commission Bulletin* 1881:264.*

———. 1884. Leech culture. *U.S. Fisheries Commission Bulletin* 4:175-176.*

Hilsenhoff, W. L. 1963. Predation by the leech, *Helobdella stagnalis*, on *Tendipes plumosus* (Diptera:Tendipedidae). *Entomological Society of America Annals* 56:252.

———. 1964. Predation by the leech, *Helobdella nepheloidea*, on larvae of *Tendipes plumosus* (Diptera:Tendipedidae). *Entomological Society of America Annals* 57:139.

Hoffman, G. L. 1967. *Parasites of North American freshwater fishes*. Berkeley and Los Angeles: University of California Press. Hirudinea, pp. 288-298.

Hoffman, Richard L. 1964. A new species of *Cystobranchus* from southwestern Virginia (Hirudinea:Piscicolidae). *American Midland Naturalist* 72(2): 309-395.

Hubbs, Carl L., and Karl F. Lagler. 1947. Fishes of the Great Lakes region. *Cranbrook Institute of Science Bulletin*, no. 26.**

Hugghins, E. J. 1958. Studies on parasites of fishes in South Dakota. *Journal of Parasitology* 44(4), sect. 2:33 (abstract).*

Hutton, Robert F., and Franklin Sogandares-Bernal. 1959. Notes on the distribution of the leech, *Myzobdella lugubris* Leidy, and its association with mortality of the blue crab, *Callinectes sapidus* Rathburn. *Journal of Parasitology* 45(4):384.

Ives, M. L. 1938. A leech and his leeches. *Natural History of New York* 42:366-370.*

Jarry, D. 1960. Une curieuse coaction de parasitisme: l'association entre *Glossiphonia complanata* et *Erpobdella octoculata*. *Terre et Vie* 107:51-55.

Judd, William W. 1968. Crayfish in the vicinity of London, Ontario. *National Museum of Canada Natural History Papers*, no. 41.

———. 1969. Studies of the Byron Bog in southwestern Ontario. XXXVII. Leeches (Hirudinea) collected in the bog. *Canadian Field-Naturalist* 83:168.

Kenk, Roman. 1946. *A bibliography of the invertebrates of Michigan (exclusive of insects)*. Ann Arbor: University of Michigan Press. Hirudinea, pp. 300-303.

———. 1949. Animal life of temporary and permanent ponds in southern Michigan. *Museum of Zoology of the University of Michigan Miscellaneous Publications*, no. 71. Hirudinea, pp. 38-39.

Kinberg, J. G. 1867. Annulata nova (continuatio). *Öfversigt af Kongl. Vetenskapsakademiens förhandlingar* (Stockholm) 23:337-357. Hirudinea, pp. 356-357.*

Kozur, Heinz. 1970. Fossile Hirudineen aus dem Oberjura von Bayern. *Lethala* 3(3):225-232.

Kraatz, W. C. 1921. A preliminary general survey of the macro-fauna of Mirror Lake on the Ohio State University campus. *Ohio Journal of Science* 21(5):137-182. Hirudinea, pp. 150-151.

Leidy, Joseph. 1852. Description of *Myzobdella*. *Academy of Natural Sciences of Philadelphia Proceedings* 1851-1853:243.

―――. 1868. Notice of some American leeches. *Academy of Natural Sciences of Philadelphia Proceedings* 20:229-230.

―――. 1870. Description of *Nephelis punctata*. *Academy of Natural Sciences of Philadelphia Proceedings* 22:89.

Linnaeus, Carl von. 1758. *Systema naturae*. 10th ed. Stockholm. Leeches, p. 649.

Livanow, N. 1902. Die Hirudineen-Gattung *Hemiclepsis* Vejd. *Zoologische Jahrbücher, Abteilung für Systematik, Geographie und Biologie der Tiere* 17:339-362.*

Lynch, D. L., et al. 1968. Characteristics of leech blood protein. *Illinois State Academy of Science Transactions* 61(3):310-312.

McAnnaly, Roy D., and Donald V. Moore. 1966. Predation by the leech *Helobdella punctato-lineata* upon *Australorbis glabratus* under laboratory conditions. *Journal of Parasitology* 52(1):196-197.

MacCallum, W. G., and G. A. MacCallum. 1918. On the anatomy of *Ozobranchus branchiatus* (Menzies). *American Museum of Natural History Bulletin* 38:395-408.*

Mann, K. H. 1953. The life history of *Erpobdella octoculata* (L.). *Journal of Animal Ecology* 22:197-207.**

―――. 1954. The anatomy of the horse leech, *Haemopis sanguisuga* (L.), with particular reference to the excretory system. *Zoological Society of London Proceedings* 124:69-88.

―――. 1955. The ecology of the British freshwater leeches. *Journal of Animal Ecology* 24:98-119.

―――. 1956. A study of the oxygen consumption of five species of leech. *Journal of Experimental Biology* 33:615-626.

―――. 1957a. The breeding, growth and age structure of a population of the leech *Helobdella stagnalis* (L.). *Journal of Animal Ecology* 26:171-177.

―――. 1957b. A study of a population of the leech *Glossiphonia complanata* (L.). *Journal of Animal Ecology* 26:99-111.

―――. 1958. Seasonal variation in the respiratory acclimatisation of the leech *Erpobdella testacea* (Sav.). *Journal of Experimental Biology* 35:314-323.**

―――. 1961a. The life history of the leech *Erpobdella testacea* and its adaptive significance. *Oikos* 12:164-169.**

―――. 1961b. *Leeches (Hirudinea): their structure, physiology, ecology and embryology*. With an appendix on the systematics of marine leeches by Prof. E. W. Knight-Jones. New York: Pergamon Press.

Mason, William T., Jr., et al. 1970. Artificial substrate sampling, macroinvertebrates in a polluted reach of the Klamath River, Oregon. *Water Pollution Control Federation Journal* 42(8, pt. 2):R315-R328.

Mathers, Carol K. 1948. The leeches of the Okoboji region. *Iowa Academy of Science Proceedings* 55:397-425.

―――. 1954. *Haemopis kingi*, new species (Hirudinea). *American Midland Naturalist* 52:460-468.

———. 1963. *Haemopis latero-maculatum*, new species (Annelida, Hirudinea). *American Midland Naturalist* 70(1):168-174.

Mayr, Ernst. 1965. *Animal species and evolution.* Cambridge, Mass.: Harvard University Press.**

Meyer, Fred P. 1969. A potential control for leeches. *Progressive Fish-Culturist* 31(3):160-163.

Meyer, Marvin C. 1937a. Leeches of southeastern Missouri. *Ohio Journal of Science* 37(4):248-251.

———. 1937b. Notes on some leeches from Ontario and Quebec. *Canadian Field-Naturalist* 51:117-119.

———. 1939a. Notes on the leeches (Piscicolidae) living on freshwater fishes of North America. *Journal of Parasitology* 25(6 suppl.):11.

———. 1939b. Demonstration of a species of marine Piscicolidae from Florida. *Journal of Parasitology* 25(6 suppl.):22.

———. 1940. A revision of the leeches (Piscicolidae) living on freshwater fishes of North America. *American Microscopical Society Transactions* 59(3):354-376.

———. 1941a. Further studies on the North American Hirudinea. *Journal of Parasitology* 27(6 suppl.):34.

———. 1941b. The rediscovery together with the morphology of the leech, *Branchellion ravenellii* (Girard, 1850). *Journal of Parasitology* 27(4):289-298.

———. 1946a. Further notes on the leeches (Piscicolidae) living on freshwater fishes of North America. *American Microscopical Society Transactions* 65:237-249.

———. 1946b. A new leech, *Piscicola salmositica* n. sp. (Piscicolidae), from steelhead trout (*Salmo gairdneri gairdneri* Richardson, 1838). *Journal of Parasitology* 32(5):467-476.

———. 1949. On the parasitism of the leech, *Piscicola salmositica* Meyer 1946. *Journal of Parasitology* 35:215.

———. 1954. The larger animal parasites of freshwater fishes of Maine. *Maine Department of Inland Fish and Game Bulletin*, no. 1.*

———. 1959. Another unusual case of erratic hirudiniasis. *Journal of Parasitology* 45(4), sect. 2:39.

———. 1968. Moore on the Hirudinea with emphasis on his type-specimens. *U.S. National Museum Proceedings* 125(3664):1-32.

Meyer, Marvin C., and R. V. Bangham. 1950. Erratic hirudiniasis in a lake trout (*Cristivomer namaycush*). *Journal of Parasitology* 36(6)2:20.

Meyer, Marvin C., and A. A. Barden. 1955. Leeches symbiotic on Arthropoda, especially decapod Crustacea. *Wasmann Journal of Biology* 13:297-311.

Meyer, Marvin C., and J. P. Moore. 1954. Notes on Canadian leeches (Hirudinea) with a description of a new species. *Wasmann Journal of Biology* 12:63-96.

Miller, John A. 1929. The leeches of Ohio. Ohio State University, *Franz Theodore Stone Laboratory Contributions*, no. 2.

———. 1933-1945. Studies in the biology of the leech. *Ohio Journal of Science*

I (1933) 33(6):460-463; II (1934) 34(1):57-61; III (1934) 34(5):318-322; IV (1936) 36(6):343-348; V (1942) 42(1):45-52; VI (1943) 43(5):198-200; VII (1944) 44(1):31-35; VIII (1944) 44(4):177-187; IX (1945) 45(6):233-246. (Titles vary.)

———. 1937. A study of the leeches of Michigan with keys to orders, suborders and species. *Ohio Journal of Science* 37:85-90.

Milner, James W. 1874. Report on the fisheries of the Great Lakes; the results of inquiries prosecuted in 1871 and 1872. *U.S. Commission of Fish and Fisheries Report*, pt. II, for 1872 and 1873. Pp. 45, 64.

Moore, J. E. 1964. Notes on the leeches (Hirudinea) of Alberta. *National Museum of Canada Natural History Papers*, no. 27.

———. 1966a. Further notes on Alberta leeches (Hirudinea). *National Museum of Canada Natural History Papers*, no. 32.

———. 1966b. New records of leeches (Hirudinea) for Saskatchewan. *Canadian Field-Naturalist* 80(1):59-60.

Moore, J. Percy. 1898. The leeches of the U.S. National Museum. *U.S. National Museum Proceedings* 21(1160):543-563.

———. 1900. A description of *Microbdella biannulata* with especial regard to the constitution of the leech somite. *Academy of Natural Sciences of Philadelphia Proceedings* 5:50-73.

———. 1901. The Hirudinea of Illinois. *Illinois State Laboratory of Natural History Bulletin* 5:479-547.

———. 1906. Hirudinea and Oligochaeta collected in the Great Lakes region. *Bureau of Fisheries Bulletin* 25(598):153-172 (1905).

———. 1911. Hirudinea of southern Patagonia. *Princeton University Expeditions to Patagonia Reports, 1896-1899* 3(Zoology, pt. 7):669-687.

———. 1912. Leeches of Minnesota. In Geological and Natural History Survey of Minnesota, *Zoology* 5(pt. III, Classification):64-150.

———. 1918. The leeches (Hirudinea). In H. B. Ward and G. C. Whipple, eds., *Fresh water biology*. New York: John Wiley. Pp. 646-660.

———. 1920. The leeches. Lake Maxinkuckee, a physical and biological survey. Indiana Department of Conservation 2:87-95.

———. 1922. The fresh-water leeches (Hirudinea) of southern Canada. *Canadian Field-Naturalist* 36:6-11, 37-39.

———. 1923. The control of blood-sucking leeches, with an account of the leeches of Palisades Interstate Park. New York State College of Forestry, *Roosevelt Wild Life Bulletin* 2(1):1-53.

———. 1924a. The anatomy and systematic position of the Chilean terrestrial leech, *Cardea valdiviana* (Philippi). *Academy of Natural Sciences of Philadelphia Proceedings* 76:29-48.

———. 1924b. The leeches (Hirudinea) of Lake Nipigon. *University of Toronto Studies in Biology* 23:17-30. Duplicated article: *University of Pennsylvania Zoological Laboratory Contributions*, 1924-1925, vol. 24, no. 21.

———. 1936. The leeches of Lake Nipissing. *Canadian Field-Naturalist* 50:112-114.

———. 1939. *Helobdella punctato-lineata*, a new leech from Puerto Rico. *Puerto Rico Journal of Public Health and Tropical Medicine* 1939:422-428.

———. 1946. The anatomy and systematic position of *Myzobdella lugubris* (Leidy). *Academy of Natural Sciences of Philadelphia Notulae Naturae*, no. 184.

———. 1949. (Description of *Dina bucera*.) In Kenk (1949), p. 38.

———. 1952. Professor Verrill's freshwater leeches. *Academy of Natural Sciences of Philadelphia Notulae Naturae*, no. 245.

———. 1953. Three undescribed North American leeches (Hirudinea). *Academy of Natural Sciences of Philadelphia Notulae Naturae*, no. 250.

———. 1959. Hirudinea. In W. T. Edmondson, ed., *Fresh-water biology*. 2nd ed. New York: John Wiley. Pp. 542-557.

Moore, J. Percy, and Marvin C. Meyer. 1951. Leeches (Hirudinea) from Alaska and adjacent waters. *Wasmann Journal of Biology* 9:11-77.

Mozley, Alan. 1932. A biological study of a temporary pond in western Canada. *American Naturalist* 66:235-249.

Mullin, Catharine Agnes. 1925. Some observations on the habits of leeches. *Iowa Academy of Science Proceedings* 32:415-417.*

———. 1926a. Study of the leeches of the Okoboji Lake region. Ph.D. thesis, University of Iowa.

———. 1926b. Study of the leeches of the Okoboji Lake region. *Anatomical Record* 34:164.*

Mundie, J. H. 1959. The diurnal activity of the larger invertebrates at the surface of Lac la Ronge, Saskatchewan. *Canadian Journal of Zoology* 37: 945-956.*

Muttkowski, Richard A. 1918. The fauna of Lake Mendota: a qualitative and quantitative survey with special reference to the insects. *Wisconsin Academy of Sciences, Arts and Letters Transactions* 19(1):374-482. Hirudinea, pp. 391-392.

Myers, Raymond J. 1932. Histological changes in the body of *Placobdella parasitica* associated with hypodermic impregnation. *Anatomical Record* 54, suppl. 88.*

———. 1935. Behavior and morphological changes in the leech *Placobdella parasitica* during hypodermic insemination. *Journal of Morphology* 53(3): 617-653.

Nachtreib, Henry F., Ernest E. Hemingway, and J. Percy Moore. 1912. The leeches of Minnesota. Geological and Natural History Survey of Minnesota, *Zoology*, vol. 5.

Nicholson, Henry Alleyne. 1872. Preliminary report on dredgings in Lake Ontario. *Annals and Magazine of Natural History*, ser. 4, 10:276-285.

———. 1873. Contributions to a Fauna Canadensis, being an account of the animals dredged in Lake Ontario in 1872. *Canadian Journal of Science, Literature and History* 13:493-498.*

Nigrelli, R. F. 1929. On the cytology and lifehistory of *Trypanosoma diemyctyli* and the polynuclear count of infected newts (*Triturus viridescens*). *American Microscopical Society Transactions* 48:366-387.*

———. 1946. Studies on the marine resources of southern New England. V. Parasites and diseases of the ocean pout, *Macrozoarces americanus*. II.

Platybdella buccalis sp. nov., an ichthyobdellid leech from the mouth. Yale University, *Bingham Oceanographic Collection Bulletin* 9:215-218.

Oliver, D. R. 1958. The leeches (Hirudinea) of Saskatchewan. *Canadian Field-Naturalist* 72:161-165.

Paloumpis, Andreas A., and William C. Starrett. 1960. An ecological study of benthic organisms in three Illinois River flood plain lakes. *American Midland Naturalist* 64(2):406-435.

Patrick, Ruth, et al. 1966. An ecosystematic study of the fauna and flora of the Savannah River. *Academy of Natural Sciences of Philadelphia Proceedings* 118(5):109-407.

Pawlowski, L. K. 1948. Contribution à la connaissance des Sangsues (Hirudinea) de la Nouvelle-Ecosse, de Terre-Nueve et des iles françaises Saint-Pierre et Miquelon. *Fragmenta Faunistica Musei Zoologici Polonici* 5(20): 317-353.

―――. 1955. Revision des genres *Erpobdella* de Blainville et *Dina* Blanchard (Hirudinea). *La Société des Sciences et des Lettres de Lódz Bulletin*, no. 6.

Pearse, A. S. 1924. The parasites of lake fishes. *Wisconsin Academy of Sciences, Arts and Letters Transactions* 26:437-440.

―――. 1936. Estuarine animals at Beaufort, North Carolina. *Elisha Mitchell Scientific Society Journal* 52:174-222. Hirudinea, p. 181.

Pennak, Robert W. 1953. *Fresh-water invertebrates of the United States*. New York: Ronald Press. Hirudinea, pp. 302-320.

Pinto, Cesar. 1923. Ensaio monographico dos Hirudineos. *Revista do Museu Paulista* 13:853-1118.

Pratt, H. S. 1925. *A manual of the common invertebrate animals*. Chicago: A. C. McClury and Co. Hirudinea, pp. 315-321.

Quebec Game and Fisheries Department. 1948. Control of leeches. In *Sixth Annual Report, Biological Bureau*. Montreal. Pp. 85-87.*

Rafinesque, C. S. 1820. *Annals of nature; or, annual synopsis of new genera and species of animals, plants, etc., discovered in North America*. Lexington. Hirudinea, p. 10.*

Rathbun, Richard. 1884. The leeches. In G. Brown Goode, *The fisheries and fishery industries of the United States*. Sect. I, Text, "Natural History of Useful Aquatic Animals." Washington, D.C.: U.S. Commission of Fish and Fisheries. Pp. 833-837.

Rawson, D. S. 1930. The bottom fauna of Lake Simcoe and its role in the ecology in this lake. *University of Toronto Studies in Biology* 34(40):1-183. Hirudinea, pp. 35-36.

―――. 1953. The bottom fauna of Great Slave Lake. *Fisheries Research Board of Canada Journal* 10(8):486-520.*

Rawson, D. S., and J. E. Moore. 1944. The saline lakes of Saskatchewan. *Canadian Journal of Research*, sect. D (Zoology) 22:141-201.*

Richardson, Laurence R. 1942. Observations on the migratory behavior of leeches. *Canadian Field-Naturalist* 56:67-70.

―――. 1943. The fresh-water leeches of Prince Edward Island and the problem of the distribution of leeches. *Canadian Field-Naturalist* 57:89-91.

———. 1948. *Piscicola punctata* (Verrill) feeding on the eggs of *Leucosomus corporalis* (Mitchill). *Canadian Field-Naturalist* 62:121-122.

———. 1949. The occurrence of the leech *Batrachobdella picta* (Verrill) in the dorsal subcutaneous lymph spaces of *Rana catesbiana*. *Canadian Field-Naturalist* 63(2):85-86.

———. 1969. A contribution to the systematics of the hirudinid leeches, with description of new families, genera and species. *Acta Zoologica Academiae Scientiarum Hungaricae* 15(1/2):97-149.

Richardson, Robert E. 1925a. Changes in the small bottom fauna of Peoria Lake, 1920-1922. *Illinois Natural History Survey Bulletin* 15:327-388.

———. 1925b. Illinois River bottom fauna in 1923. *Illinois Natural History Survey Bulletin* 15:391-423.

———. 1928. The bottom fauna of the middle Illinois River, 1913-1925. *Illinois Natural History Survey Bulletin* 17(12):387-475.

Ringuelet, Raúl. 1943. Sobre le morfologia y varabilidad de *Helobdella triserialis* (Em.Bl.) (Hirudinea, Glossiphoniidae). *Museo de La Plata Notas* (Buenos Aires) 8 (Zoology) (69):215-240.

———. 1944a. Sinopsis sistematica y zoogeografica de los Hirudineos de la Argentina, Chile, Paraguay y Uruguay. *Revista del Museo de La Plata*, n.s. 3 (Zoology):163-232.

———. 1944b. Revisión de los Hirudineos argentinos de los géneros *Helobdella* R.Bl., *Bactacobdella* Vig., *Cylicobdella* Gr. y *Semiscolex* Kinb. *Revista del Museo de La Plata*, n.s. 4 (Zoology) (25):6-50.

———. 1945. Hirudineos del Museo de La Plata. *Museo de La Plata Notas* 4:95-137.

Rollins, W. H. 1880. Leeches on a turtle. *American Naturalist* 14:896.*

Ruedemann, R. 1901. Hudson River beds near Albany and their taxonomic equivalents. *New York Museum Bulletin* (Albany) 8(42):487-587. Questionable fossil Hirudinea, pp. 520, 574.

Rupp, R. S., and Marvin C. Meyer. 1954. Mortality among brook trout, *Salvelinus fontinalis*, resulting from attacks of freshwater leeches. *Copeia* 1954(4):294-295.

Ruthven, A. G. 1906. An ecological survey in the Porcupine Mountains and Isle Royale, Michigan. In Michigan Geological Survey, *Annual Report 1905*. Pp. 17-55.

Ryerson, C. G. S. 1915. Notes on the Hirudinea of Georgian Bay. In Canadian Biological Board, *Contributions to Canadian Biology, 1911-1914*. Fasc. 2, pp. 165-176.

Sager, A. 1878. Notes on the Hirudinea observed in Michigan. *Essex Institute Bulletin* 9:73-76.

Sanjeeva Raj, P. J., and L. R. Penner. 1962. Concerning *Ozobranchus branchiatus* (Menzies, 1791) (Piscicolidae:Hirudinea) from Florida and Sarawak. *American Microscopical Society Transactions* 81:364-371.

Sapkarev, J. A. 1968. The taxonomy and ecology of leeches (Hirudinea) of Lake Mendota, Wisconsin. *Wisconsin Academy of Sciences, Arts and Letters Transactions* 56:225-253.

Sawyer, Roy T. 1967. The leeches of Louisiana, with notes on some North American species. *Louisiana Academy of Science Proceedings* 30:32-38.

———. 1968. Notes on the natural history of the leeches (Hirudinea) on the George Reserve, Michigan. *Ohio Journal of Science* 68(4):226-228.

———. 1969. Studies on the Hirudinea. Ph.D. thesis, University of Wales, Swansea.

———. 1970a. Observations on the natural history and behavior of *Erpobdella punctata* (Annelida:Hirudinea). *American Midland Naturalist* 83(1):65-80.

———. 1970b. The juvenile anatomy and post-hatching development of the marine leech, *Oceanobdella blennii* (Knight-Jones, 1940). *Journal of Natural History* 4:175-188.

———. 1971a. The phylogenetic development of brooding behaviour in the Hirudinea. *Hydrobiologia* 37(2):197-204.

———. 1971b. The rediscovery of the bi-annulate leech, *Oligobdella biannulata* (Moore, 1900), in the mountain streams of South Carolina (Annelida: Hirudinea). *Association of Southeastern Biologists Bulletin* 18(2):54.

———. 1971c. Erpobdellid leeches as new hosts for the nematomorph, *Gordius* sp. *Journal of Parasitology* 57(2):285.

Say, Thomas. 1824. On *Hirudo parasitica, lateralis, marmorata* and *decora*, on the voyage of Major Long. In W. H. Keating, *Narrative of the expedition to the source of St. Peters River, Lake Winnipeck, Lake of the Woods in 1823 under Stephen H. Long*, 2 vols. Philadelphia. Appendix of natural history. Zoology, 2:253; Vermes, 2:266-268.

Schmidt, Gerald D., and Kathleen Chaloupka. 1969. *Alloglossidium hirudicola* sp. n., a neotenic trematode (Plagiorchiidae) from leeches, *Haemopis* sp. *Journal of Parasitology* 55(6):1185-86.

Scudder, G. G. E., and K. H. Mann. 1968. The leeches of some lakes in the southern interior plateau region of British Columbia. *Syesis* 1(1/2):203-209.

Smith, R. I. 1942. Nervous control of chromatophores in the leech *Placobdella parasitica*. *Physiological Zoology* (Chicago) 15:410-417.

Smith, S. J. 1871. (Hirudinea.) In *Report to the Secretary of War* 1871(2): 1024-26.

———. 1874. Sketch of the invertebrate fauna of Lake Superior. In *U.S. Fisheries Commission Report for 1872-73*. Pt. 2, pp. 690-707.

Smith, S. J., and A. E. Verrill. 1871. Notice of the Invertebrata dredged in Lake Superior in 1871 (worms). *American Journal of Science and Arts*, ser. 3, 2:374, 448-454.

Soós, A. 1963. Identification key to the species of the genus *Dina* R. Blanchard, 1892 (emend. Mann, 1952). *Acta Universitatis Szegediensis, Acta Biologica*, n.s. 9:253-261.

———. 1965. Identification key to the leech (Hirudinoidea) genera of the world, with a catalogue of the species. I. Family: Piscicolidae. *Acta Zoologica Academiae Scientiarum Hungaricae* 11(3/4):417-463.

———. 1966a. Identification key to the leech (Hirudinoidea) genera of the world, with a catalogue of the species. II. Families: Semiscolecidae, Trematobdellidae, Americobdellidae, Diestecostomatidae. *Acta Zoologica Academiae Scientiarum Hungaricae* 12(1/2):145-160.

———. 1966b. Identification key to the leech (Hirudinoidea) genera of the world, with a catalogue of the species. III. Family: Erpobdellidae. *Acta Zoologica Academiae Scientiarum Hungaricae* 12(3/4):371-407.

———. 1966c. On the genus *Glossiphonia* Johnson, 1916, with a key and catalogue to the species (Hirudinoidea:Glossiphoniidae). *Annales Historico-Naturales Musei Nationalis Hungarici* 58:271-279.

———. 1967. On the genus *Batracobdella* Viguier, 1879, with a key and catalogue to the species. *Annales Historico-Naturales Musei Nationalis Hungarici* 59:243-257.

———. 1969. Identification key to the leech (Hirudinoidea) genera of the world, with a catalogue of the species. V. Family: Hirudinidae. *Acta Zoologica Academiae Scientiarum Hungaricae* 15(1/2):151-201.

Sooter, C. A. 1937. Leech infesting young waterfowl in north-west Iowa. *Journal of Parasitology* 23(1):108-109.

Thomas, Allan E. 1969. Mortality due to leech infestations in an incubation channel. *Progressive Fish-Culturist* 31(3):164-165.

Thomas, M. L. H. 1966. Benthos of four Lake Superior bays. *Canadian Field-Naturalist* 80(4):200-212.

Thompson, David H. 1927. An epidemic of leeches on fishes in Rock River. *Illinois State Natural History Survey Bulletin* 17:195-201.

Thut, Rudolph N. 1969. A study of the profundal bottom fauna of Lake Washington. *Ecological Monographs* 39:79-100.

Townes, H. K., Jr. 1937. A biological survey of the Allegheny and Chemung water sheds. VI. Studies on the food organisms of fish. *New York Conservation Department Biological Survey* 12:162-175.

Van Harreveld, A., F. I. Khattab, and Jana Steiner. 1969. Extracellular spaces of the central nervous system of the leech *Mooreobdella fervida*. *Journal of Neurobiology* 1(1):23-40.

Venard, Carl E. 1940. Studies on parasites of Reelfoot Lake fish. I. Parasites of the largemouthed black bass, *Huro salmoides* (Lacépéde). Tennessee Academy of Science, *Reelfoot Lake Biological Station Report* 4:43-63.*

———. 1941. Studies on parasites of Reelfoot Lake fish. II. Parasites of the warmouth bass, *Chaenobryttus gulosus* (Cuvier and Valenciennes). Tennessee Academy of Science, *Reelfoot Lake Biological Station Report* 5:14-16.*

Verrill, Addison Emory. 1872a. Espéces américaines d'Hirudinées. *Journal de Zoologie* (Paris) 1:200-201.*

———. 1872b. Brief contributions to zoology from the Museum of Yale College. XVII. Descriptions of North American freshwater leeches. *American Journal of Science and Arts*, ser. 3, 3(14):126-139.

———. 1873a. Report upon the invertebrate animals of Vineyard Sound and the adjacent waters, with an account of the physical characters of the region. In *U.S. Fisheries Commission Report for 1871-72*. Pt. 1, pp. 458-460, 624-626.

———. 1873b. Contributions to a Fauna Canadensis, being an account of the animals dredged in Lake Ontario in 1872 by H. Alleyne Nicholson. *American Journal of Science and Arts*, ser. 3, 5:387-389.*

———. 1874a. Synopsis of North American freshwater leeches. In *U.S. Fisheries Commission Report for 1872-73*. Pt. 2, pp. 666-689.

———. 1874b. List of leeches collected by Hayden's expedition in Colorado, 1873. In *U.S. Geological and Geographical Survey Annual Report* 7:623.

———. 1875a. Report upon the collections of freshwater leeches made in portions of Nevada, Utah, Colorado, New Mexico and Arizona, during the years 1872, 1873 and 1874. In *U.S. Geological and Geographical Survey Report, west of the 100th meridian* 5:955-967.

———. 1875b. Results of dredging expeditions of the New England coast in 1874. *American Journal of Science and Arts*, ser. 3, 10:32-36, 196-198.*

Viosca, P. 1962. Observations on the biology of the leech *Philobdella gracile* Moore in southeastern Louisiana. *Tulane Studies in Zoology* 9:243-244.

Wales, J. H., and H. Wolf. 1955. Three protozoan diseases of trout in California. *California Fish and Game* 41:183-187.*

Ward, H. B. 1896. (Hirudinea.) *Michigan Fisheries Commission Bulletin* 6:13.

———. 1902. Notes on the leeches of Nebraska. In Nebraska State Board of Agriculture, *Annual Report*. Lincoln, Nebr., 1901. Pp. 267-280.

Weber, Maurice. 1913. Hirudinées colombiennes. Voyage d'exploration scientifique en Colombie, Dr. O. Fuhrmann et Dr. Eug. Mayor. *La Société Neuchateloise des Sciences Naturelles Memoires* 5:731-747.

———. 1915. Monographie des Hirudinées sud-americaines. Thése présentée à la Faculté des Sciences, Neuchatel.

Webster, Edward J. 1967. An autoradiographic study of invertebrate uptake of DDT-Cl^{36}. *Ohio Journal of Science* 67(5):300-307.

Weston, Robert Spurr, and C. E. Turner. 1917. Studies on the digestion of a sewage-filter effluent by a small and otherwise unpolluted stream. Massachusetts Institute of Technology, *Contributions from the Sanitary Research Laboratory and Sewage Experimental Station* 10:1-96.*

Whitman, C. O. 1886. The leeches of Japan. *Quarterly Journal of Microscopical Science* 26:317-416.

———. 1889. Some new facts about the Hirudinea. *Journal of Morphology* 2:586-599.

———. 1891. Description of *Clepsine plana*. *Journal of Morphology* 4:409-418.

———. 1892. The metamerism of Clepsine. In *Festschrift zum 70 Gubertstage Rudolf Leuckarts*. Leipzig. Pp. 385-395.

———. 1898. Animal behavior. In Woods Hole Marine Biology Laboratory, *Biological Lectures*. Pp. 285-338.

Woo, Patrick T. H. 1969a. The development of *Trypanosoma canadensis* of *Rana pipiens* in *Placobdella* sp. *Canadian Journal of Zoology* 47(6):1257-59.

———. 1969b. The life cycle of *Trypanosoma chrysemydis*. *Canadian Journal of Zoology* 47(6):1139-51.

Wright, H. E., Jr., and David G. Frey, eds. 1965. *The Quaternary of the United States*. Princeton, N.J.: Princeton University Press.**

Yagi, K., H. A. Bern, and I. R. Hagadorn. 1963. Action potentials of neurosecretory neurons in the leech, *Theromyzon rude*. *General and Comparative Endocrinology* 3:490-495.

FIGURES

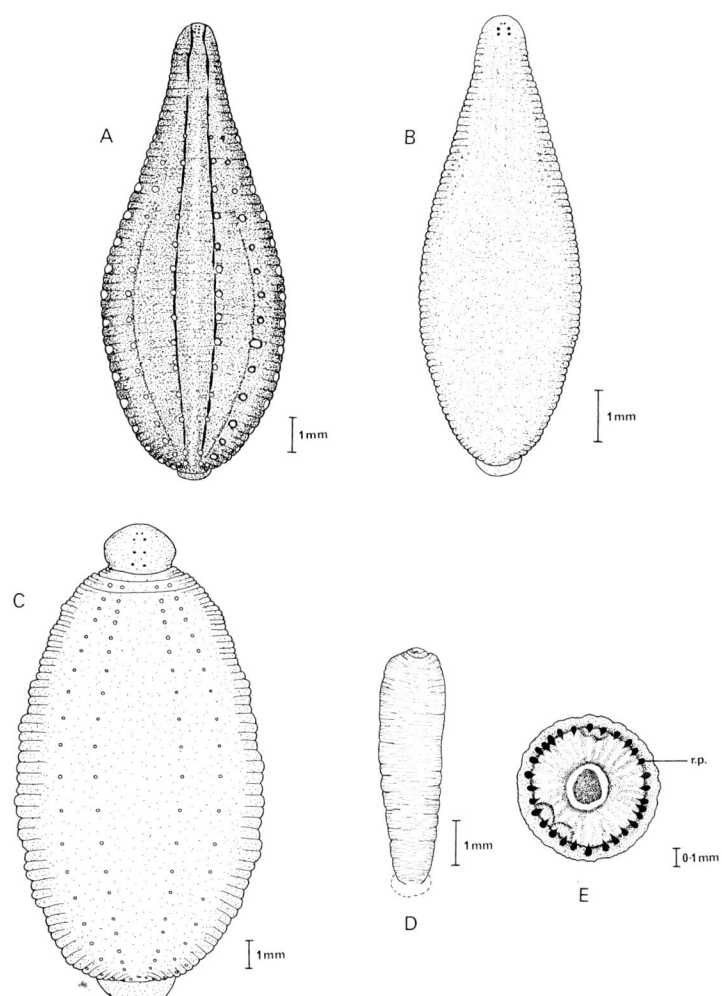

FIG. 1. A. *Glossiphonia complanata*. B. *G. heteroclita*. C. *Theromyzon meyeri*. D. *Actinobdella inequiannulata*, without caudal sucker. E. Caudal sucker of D, showing retractile papillae (r.p.) around sucker cavity rim. A-D, dorsal view; E, ventral view.

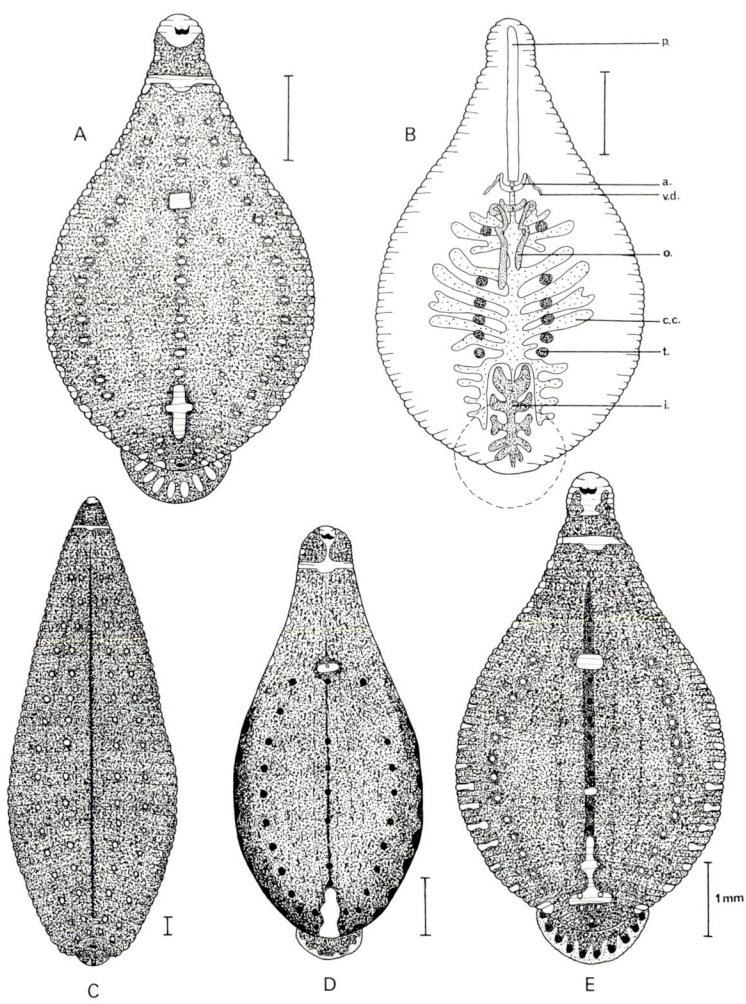

Fig. 2. *Batracobdella*. A. *B. michiganensis*. B. Digestive and reproductive systems of A. C. *B. picta*. D-E. *B. phalera*. A, C-E, dorsal view. a., atrium; c.c., crop caecum; i., intestine; o., ovary; p., pharynx; t., testis; v.d., vas deferens.

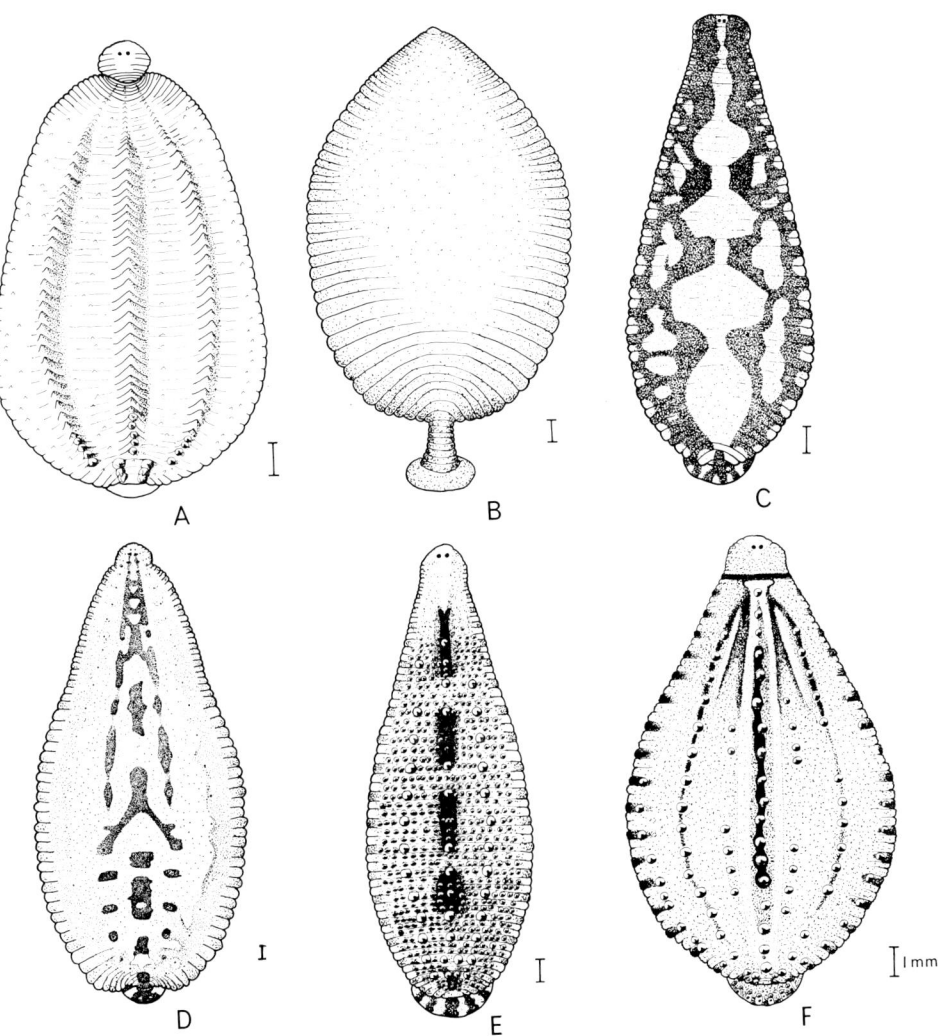

FIG. 3. *Placobdella* (dorsal view). A. *P. montifera*. B. *P. pediculata*. C. *P. parasitica*. D. *P. hollensis*. E. *P. ornata*. F. *P. papillifera*.

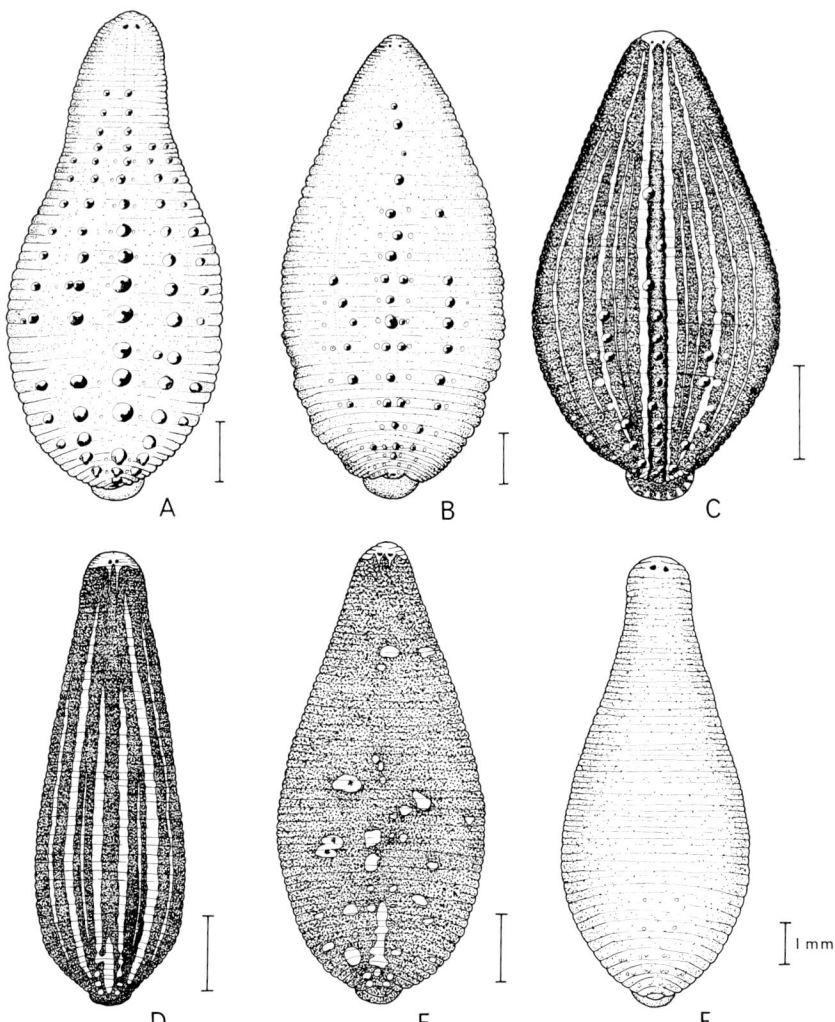

Fig. 4. *Helobdella* (dorsal view). A. *H. papillata*. B. *H. lineata*. C. *H. lineata*. D. *H. fusca*, typical form. E-F. *H. fusca*, color variants.

Fig. 5. A. *Helobdella transversa*. B. Digestive and reproductive systems of A. C. *H. stagnalis*. D. *H. elongata*. E. *Oculobdella lucida*. A, C-E, dorsal view. a., atrium; c., crop; i., intestine; p., pharynx; p.c.c., posterior crop caecum; s., scute; t., testis; v.d., vas deferens.

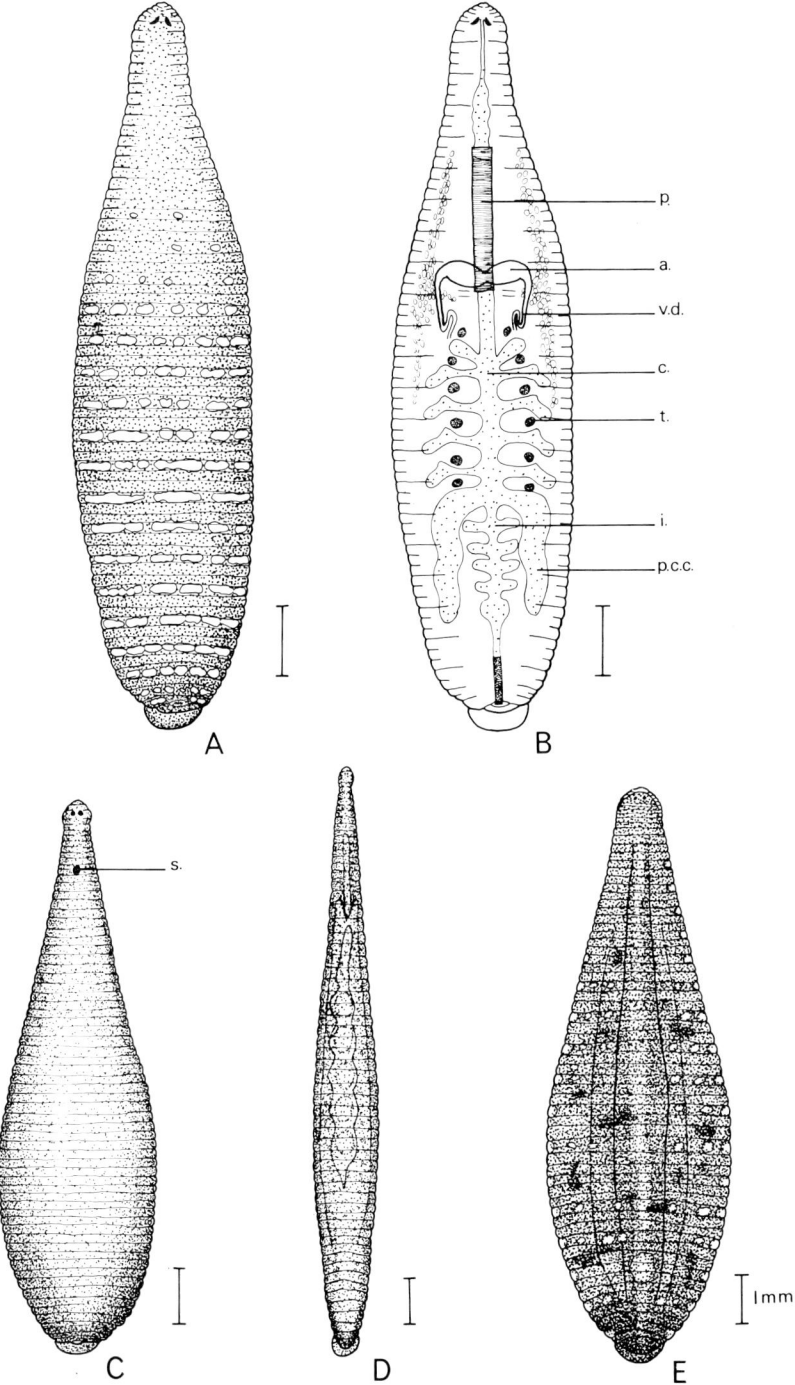

FIG. 6. A-B. *Glossiphonia complanata* from Earhardt Pond, Washtenaw County, Michigan. A. Distribution of the number of cocoons deposited by 13 individuals (average number of cocoons per individual = 6.24). B. Distribution of the number of eggs contained in 89 cocoons (average number of eggs per cocoon = 20.6). C-F. *Helobdella stagnalis*, also from Earhardt Pond. C. Distribution of the number of cocoons carried by 23 individuals (average number of cocoons per individual = 8.35). D. Distribution of the number of eggs contained in 193 cocoons (average number of eggs per cocoon = 4.23). E. Relationship between the average number of eggs per cocoon and the order in which they were laid (assuming the anterior cocoons were laid first), based on 23 individuals. F. Relationship between length and the average number of eggs per individual (open circles) and cocoons per individual (closed circles), based on 23 individuals.

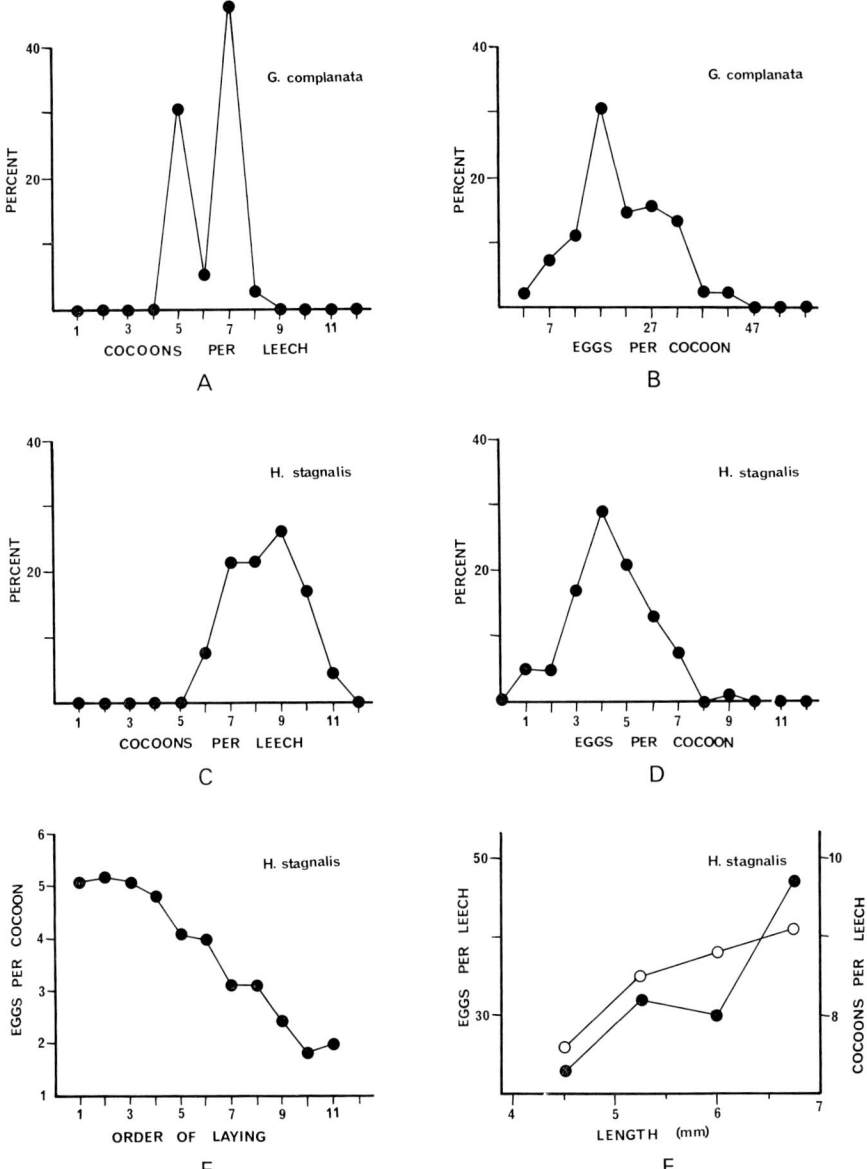

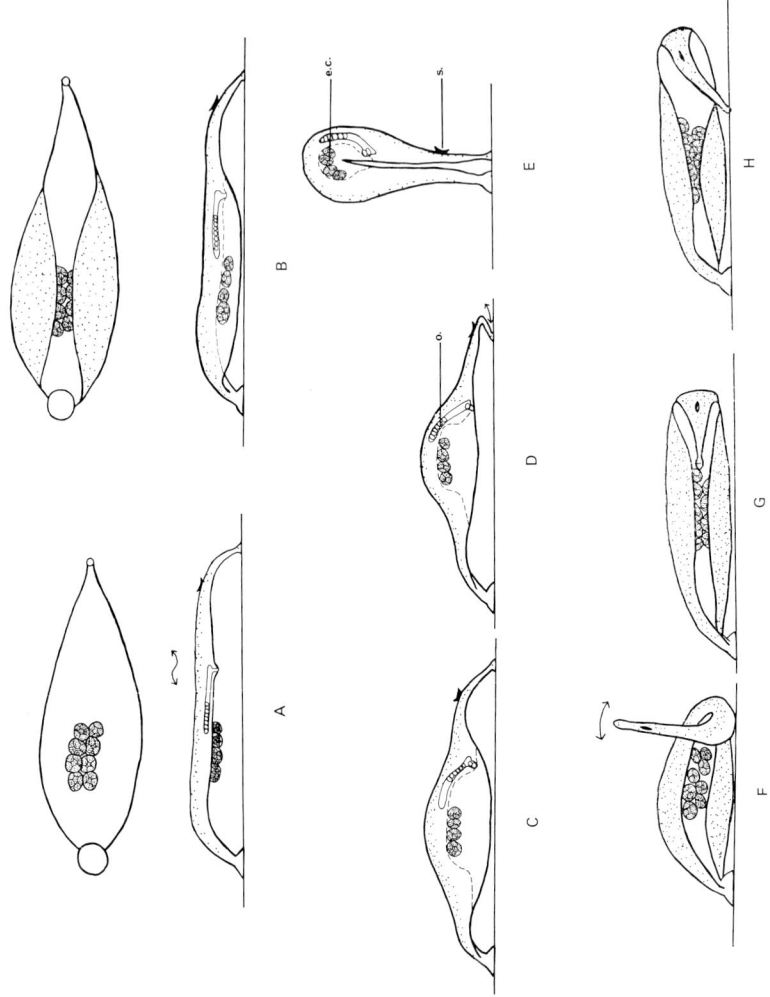

Fig. 7. Egg-laying repertoire of *Helobdella stagnalis,* described in the text. A-B, ventral and lateral views; C-H, lateral view. e.c., egg capsule; o., ovary; s., scute.

Fig. 8. Erpobdellidae (dorsal view). A. *Dina dubia.* B. *D. parva.* C. *Erpobdella punctata.* D. Three annuli showing color variants of *E. punctata.* E. *Mooreobdella microstoma.* F. *M. bucera.* G. *M. fervida.* H. *Nephelopsis obscura.*

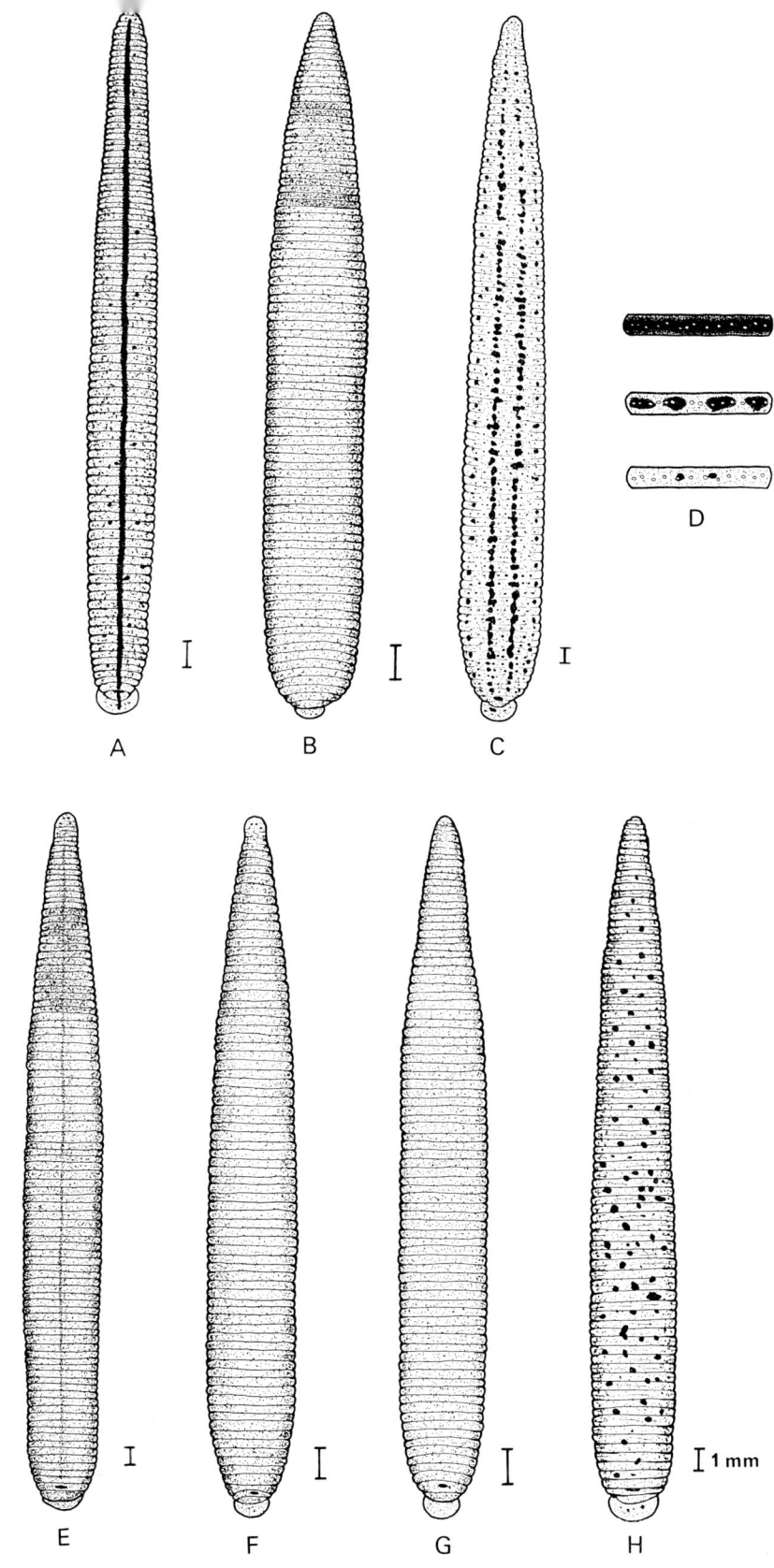

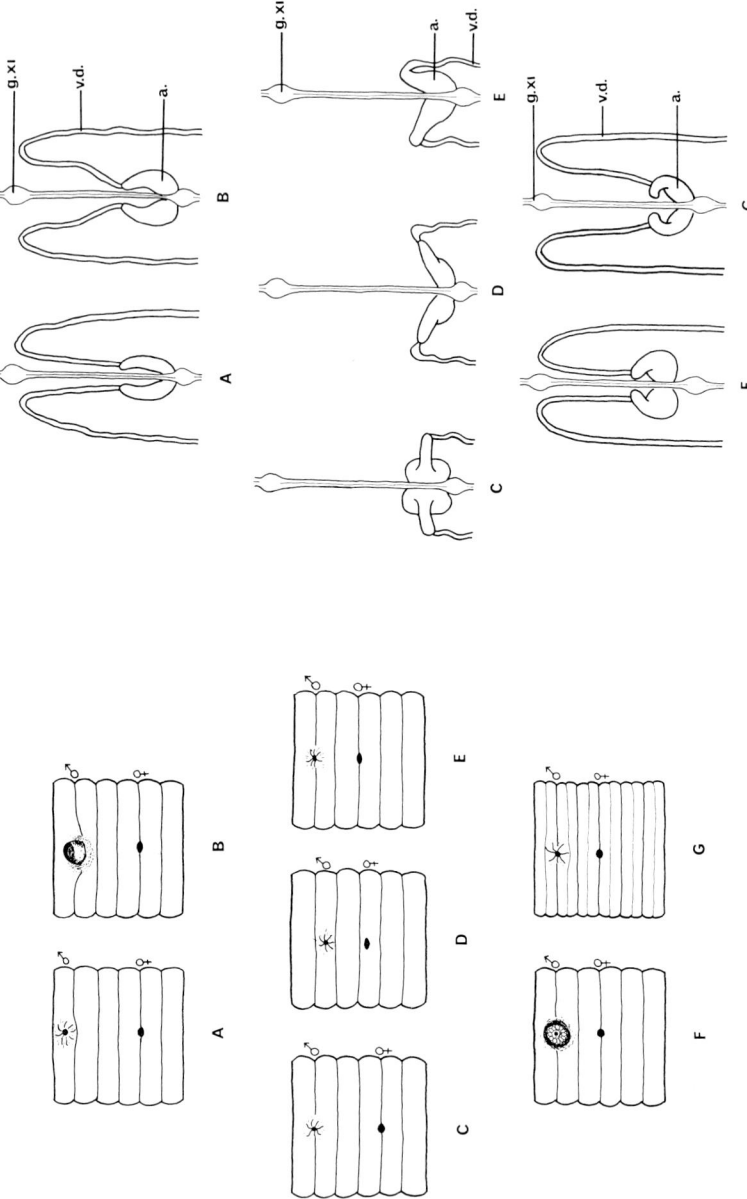

Fig. 9. Erpobdellidae. Left, ventral view showing relative positions of male and female gonopores; right, male reproductive system. A. *Dina dubia*. B. *D. parva*. C. *Mooreobdella microstoma*. D. *M. bucera*. E. *M. fervida*. F. *Erpobdella punctata*. G. *Nephelopsis obscura*. a., atrium; g., ganglion; v.d., vas deferens.

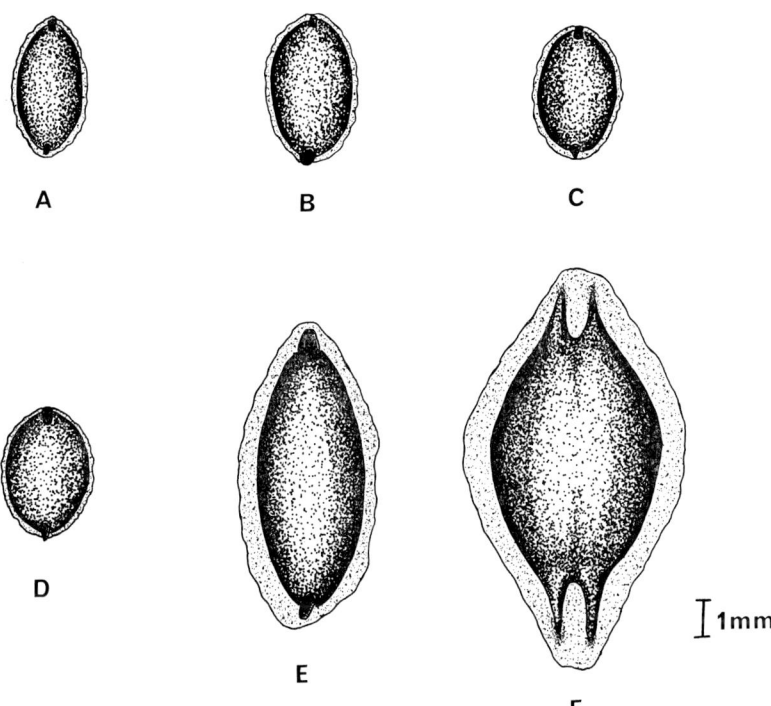

FIG. 10. Erpobdellid cocoons. A. *Mooreobdella microstoma*. B. *M. bucera*. C. *M. fervida*. D. *Dina dubia*. E. *Erpobdella punctata*. F. *Nephelopsis obscura*.

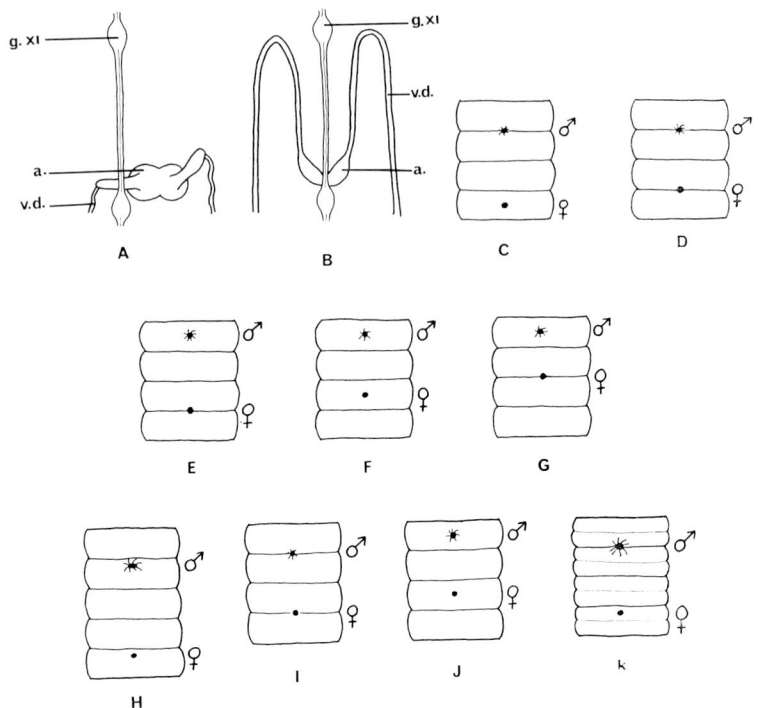

Fig. 11. Erpobdellidae. Variations of the male reproductive system (A-B) and relative positions of gonopores (C-K). A. *Mooreobdella microstoma*. B. Immature *Erpobdella punctata*. C-D. *M. bucera*. E-G. Sample of 51 individuals of *M. bucera* from Earhardt Pond, Washtenaw County, Michigan: E, 2.0%; F, 92.2%; G, 5.9%. H. *M. microstoma*. I. Immature *E. punctata*. J. *M. fervida*. K. *Nephelopsis obscura*. a., atrium; g., ganglion; v.d., vas deferens.

Fig. 12. A-C. *Mooreobdella bucera* from Earhardt Pond, Washtenaw County, Michigan. A. Relationship between the average number of eggs per cocoon and the order in which they were deposited, based on 18 isolated individuals. B. Distribution of the number of cocoons deposited by 18 isolated individuals (average number of cocoons per individual = 4.6). C. Distribution of the number of eggs contained in 103 cocoons (average number of eggs per cocoon = 5.95). D-E. *Dina dubia* from Duck Lake, Calhoun County, Michigan. D. Distribution of the number of cocoons deposited by 10 isolated individuals (average number of cocoons per individual = 7.9). E. Distribution of the number of eggs contained in 94 cocoons (average number of eggs per cocoon = 4.15).

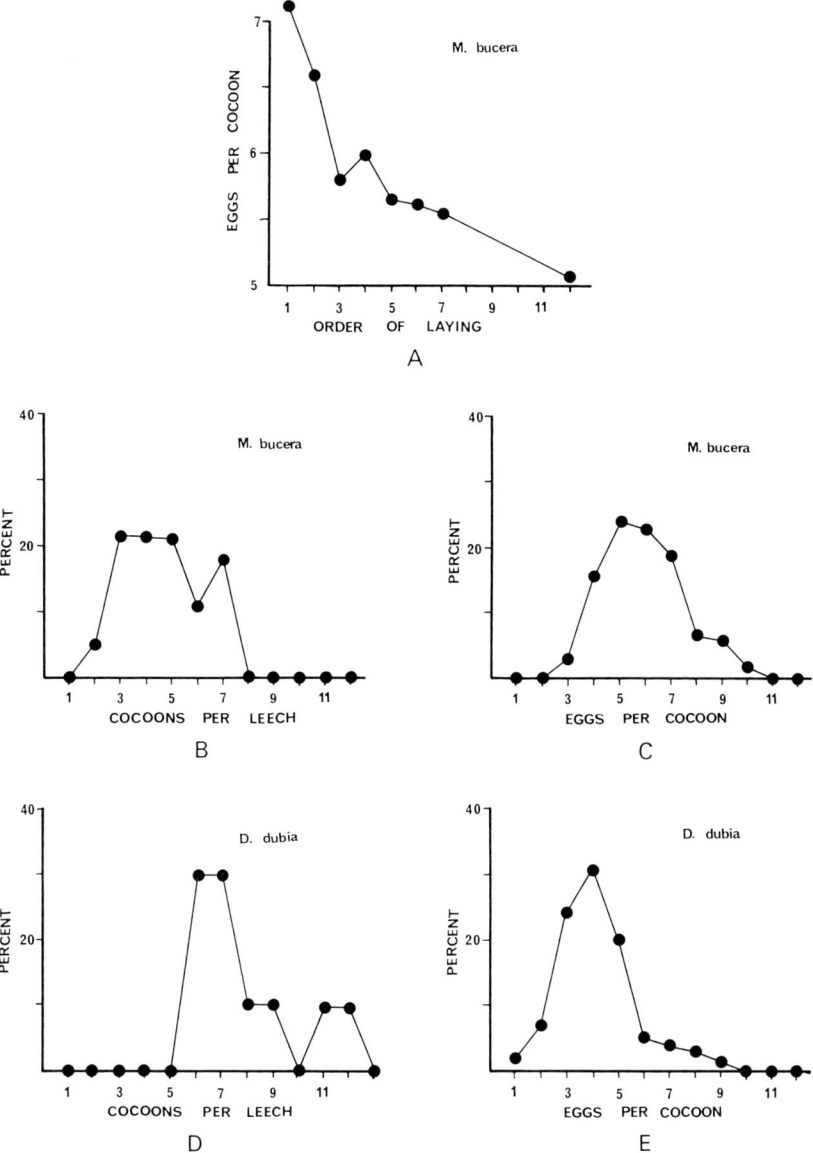

FIG. 13. Hirudinidae (dorsal view). A. *Macrobdella decora*. B. *M. ditetra* (McIntosh County, Georgia). C. *Philobdella gracilis*. D. *Haemopis marmorata*. E. *H. grandis*. F. *H. terrestris*.

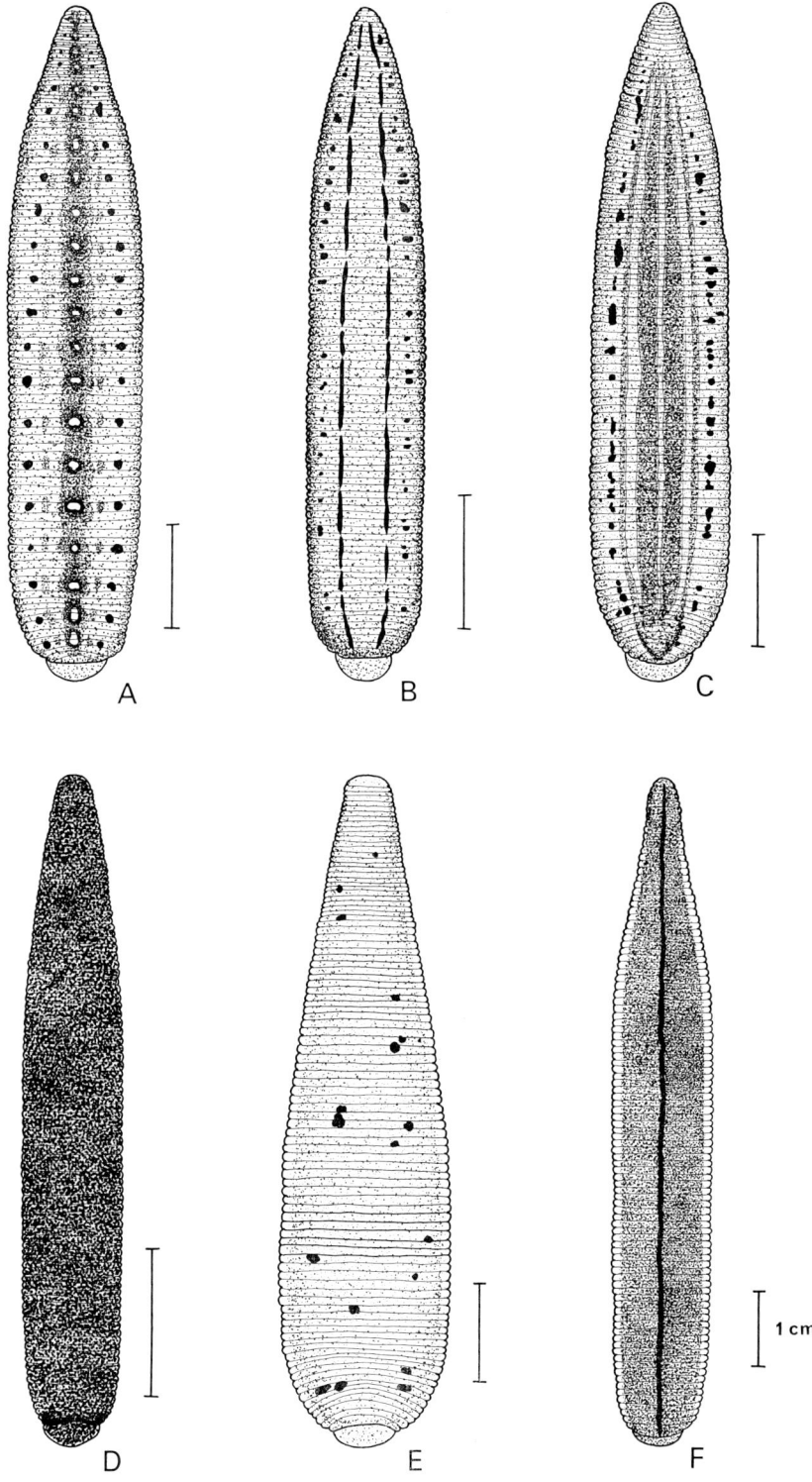

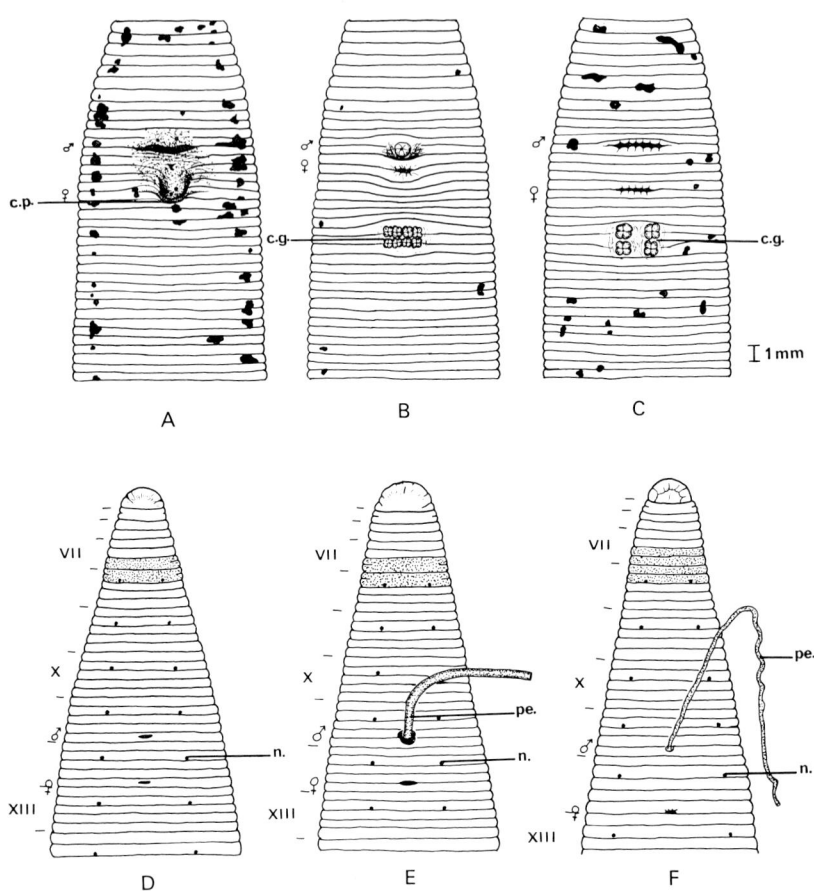

Fig. 14. Hirudinidae (ventral view). A. *Philobdella gracilis*. B. *Macrobdella ditetra* (McIntosh County, Georgia). C. *M. decora*. D. *Haemopis marmorata*. E. *H. grandis*. F. *H. terrestris*. c.g., copulatory glands; c.p., copulatory pit; n., nephridiopore; pe., penis. Shadings in D-F indicate subdivisions of annuli VIIa3 and VIIIa1.

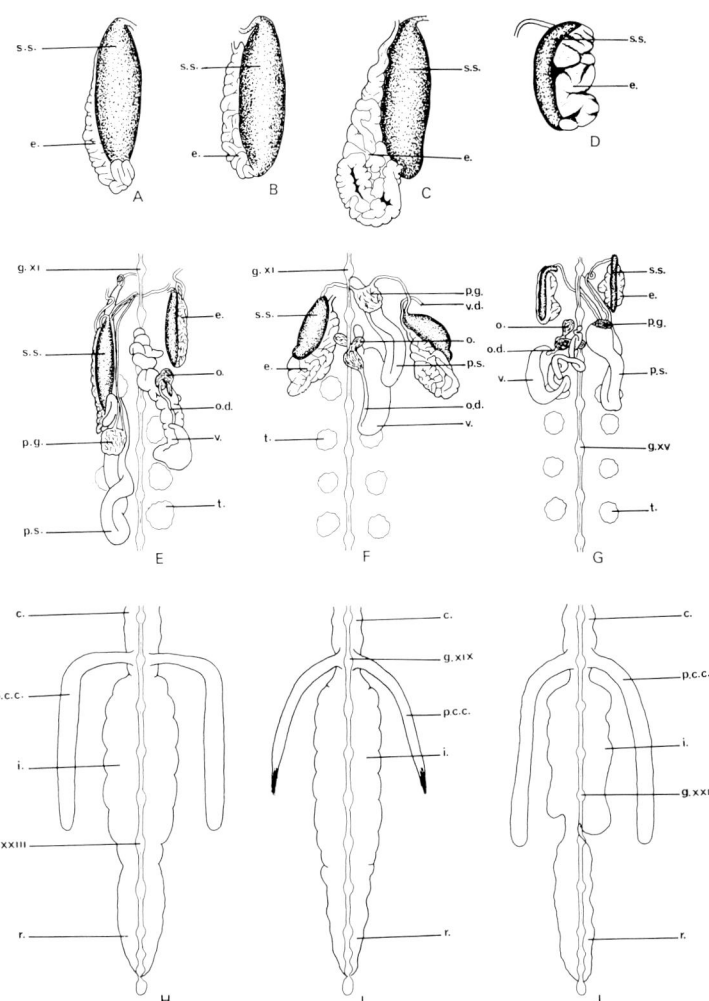

FIG. 15. *Haemopis*. A-D. Epididymis and sperm sacs. E-G. Dissections of male and female reproductive systems. H-J. Posterior part of digestive tract. A, B, E, H. *H. marmorata*. C, F, I. *H. grandis*. D, G, J. *H. terrestris*. c., crop; e., epididymis; g., ganglion; i., intestine; o., ovary; o.d., oviduct; p.c.c., posterior crop caecum; p.g., prostate gland; p.s., penis sheath; r., rectum; s.s., sperm sac; t., testis; v., vagina; v.d., vas deferens.

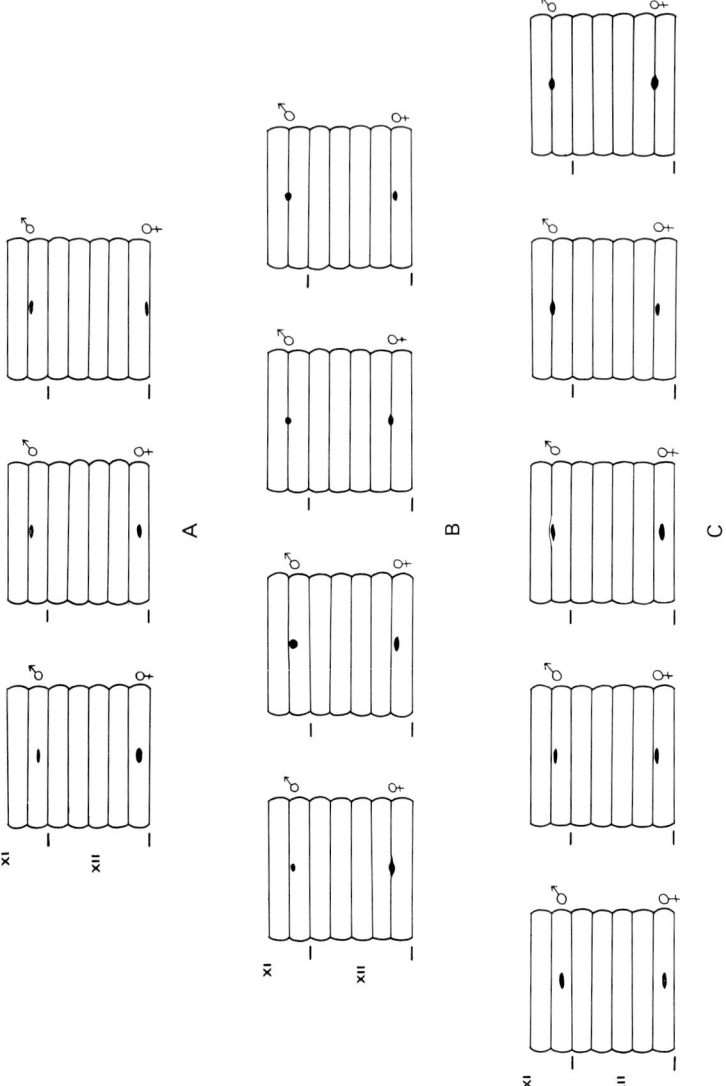

FIG. 16. Variations of the relative positions of gonopores. A. *Haemopis terrestris*. B. *H. grandis*. C. *H. marmorata*.

127

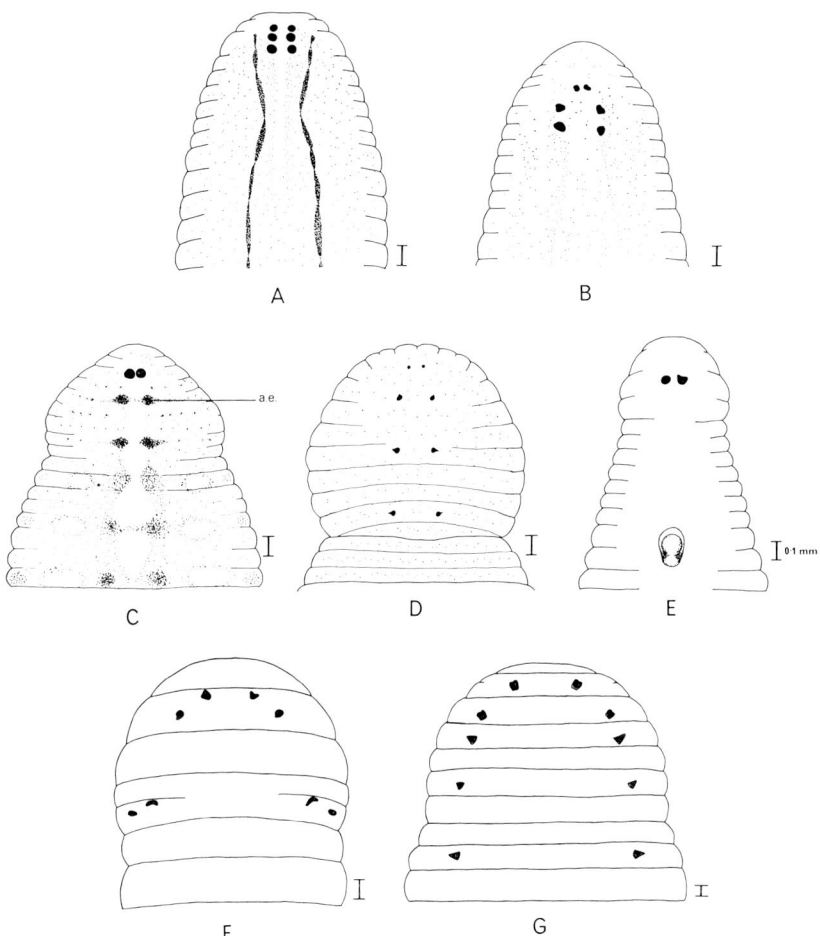

FIG. 17. Dorsal views of the head region showing various arrangements of eyes. A. *Glossiphonia complanata*. B. *G. heteroclita*. C. *Placobdella hollensis*. D. *Theromyzon meyeri*. E. *Helobdella stagnalis*. F. *Dina parva*. G. *Haemopis grandis*. a.e., accessory eyes.

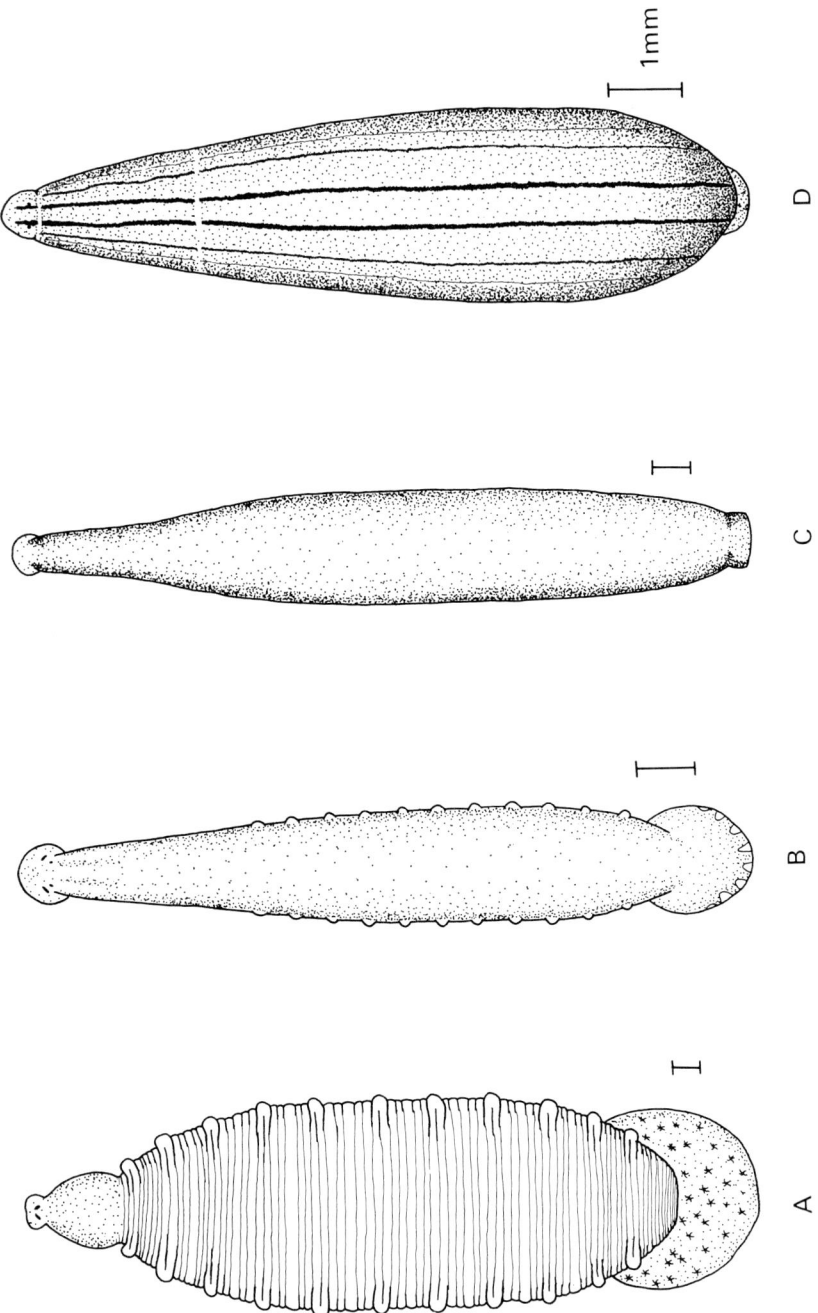

FIG. 18. Piscicolidae. A. *Cystobranchus verrilli*. B. *Piscicola punctata*. C. *Illinobdella moorei*. D. *Piscicolaria reducta*.

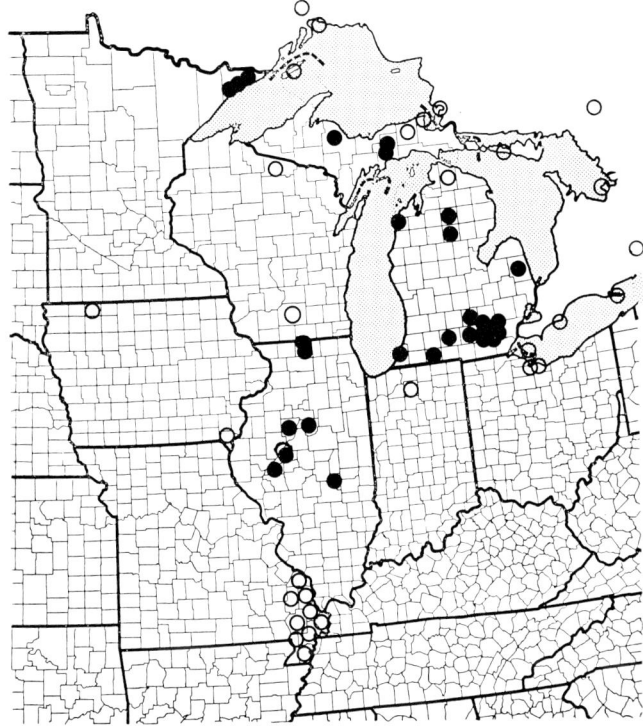

FIGS. 19-36. County outline maps of the Great Lakes region showing known positive records of the occurrence of each species; open symbols represent published records thought to be valid, and solid symbols represent new records encountered in the present study.

FIG. 19. *Glossiphonia complanata*.

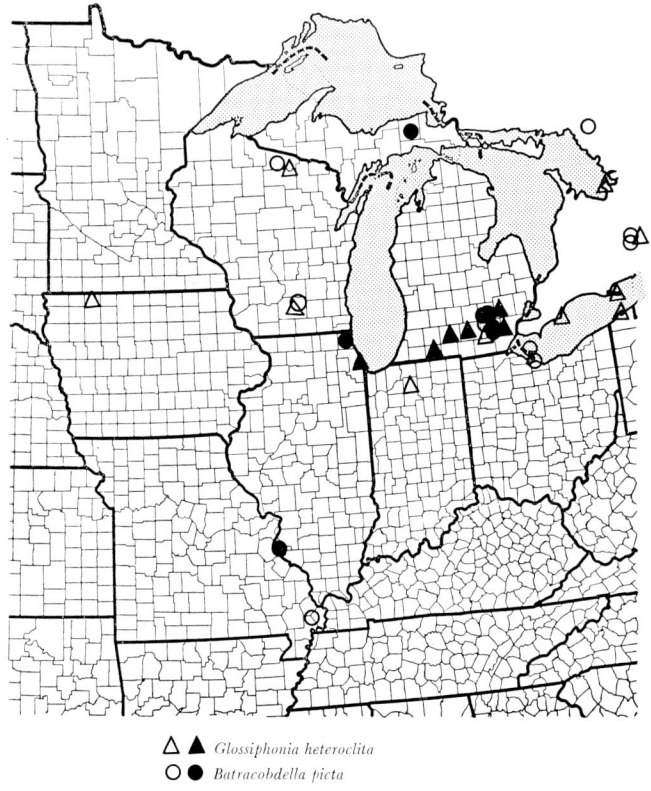

Fig. 20. *Glossiphonia heteroclita, Batracobdella picta.*

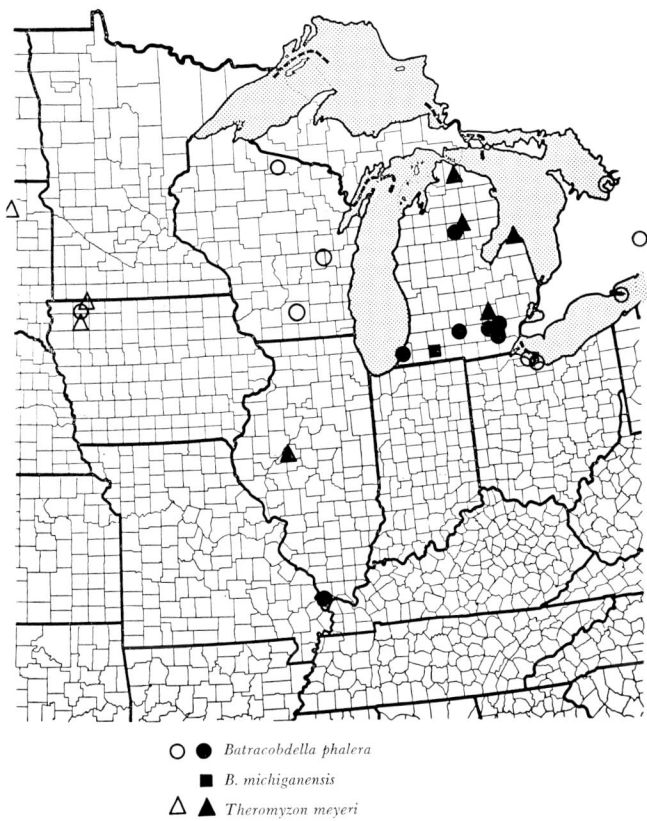

FIG. 21. *Batracobdella phalera, B. michiganensis, Theromyzon meyeri.*

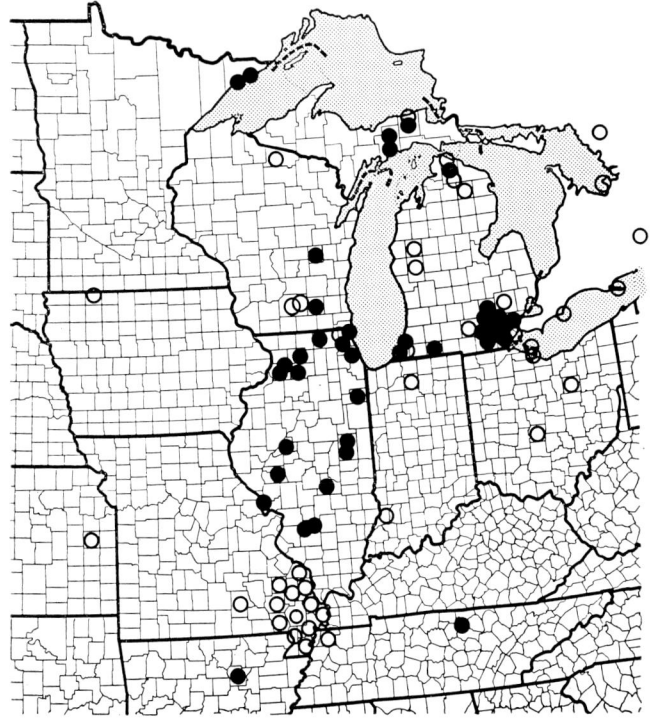

Fig. 22. *Placobdella parasitica*.

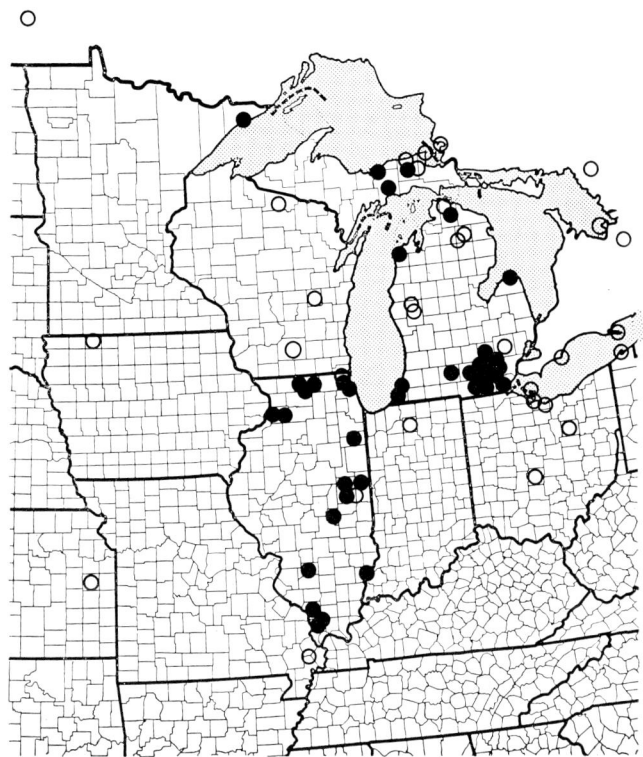

FIG. 23. *Placobdella ornata.*

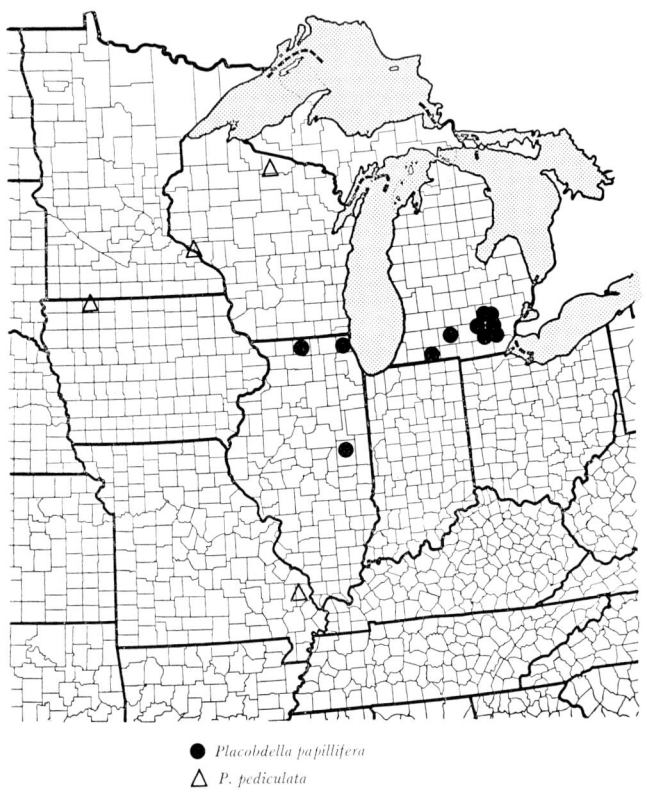

FIG. 24. *Placobdella papillifera, P. pediculata.*

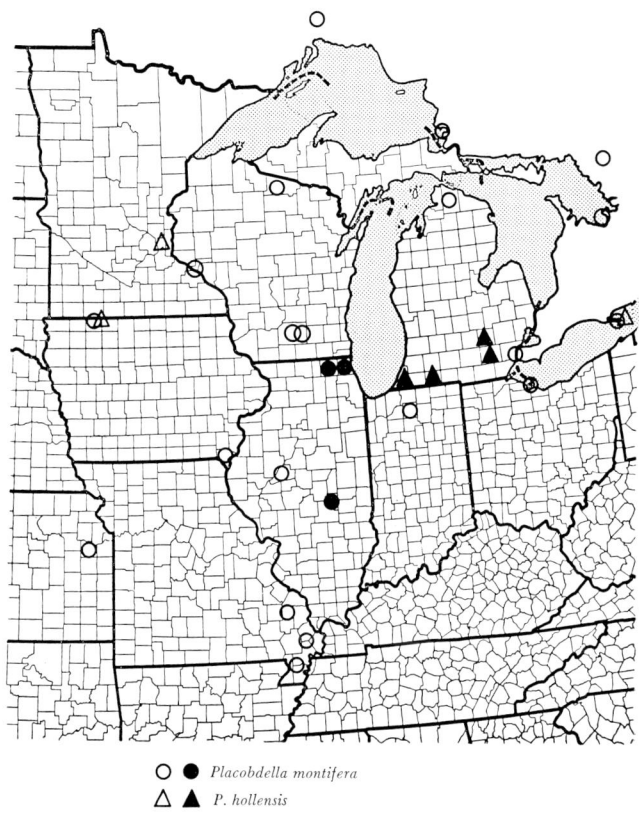

FIG. 25. *Placobdella montifera, P. hollensis*.

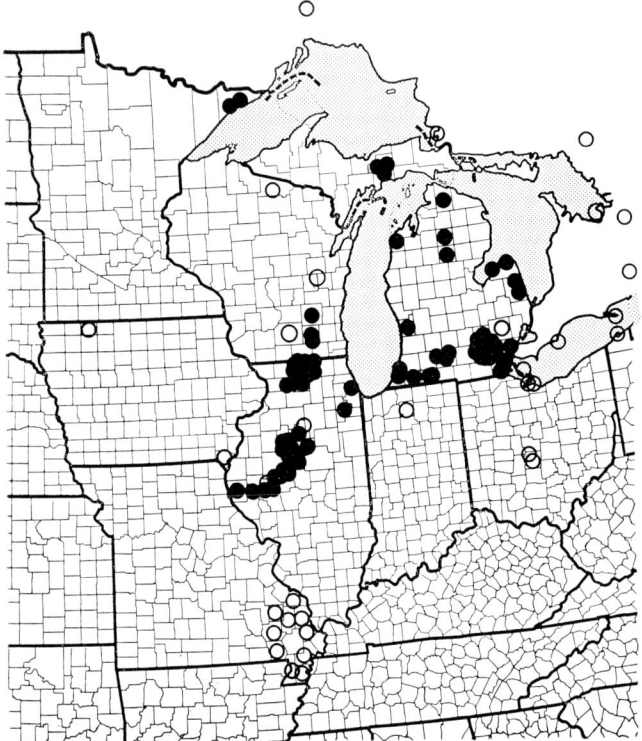

Fig. 26. *Helobdella stagnalis*.

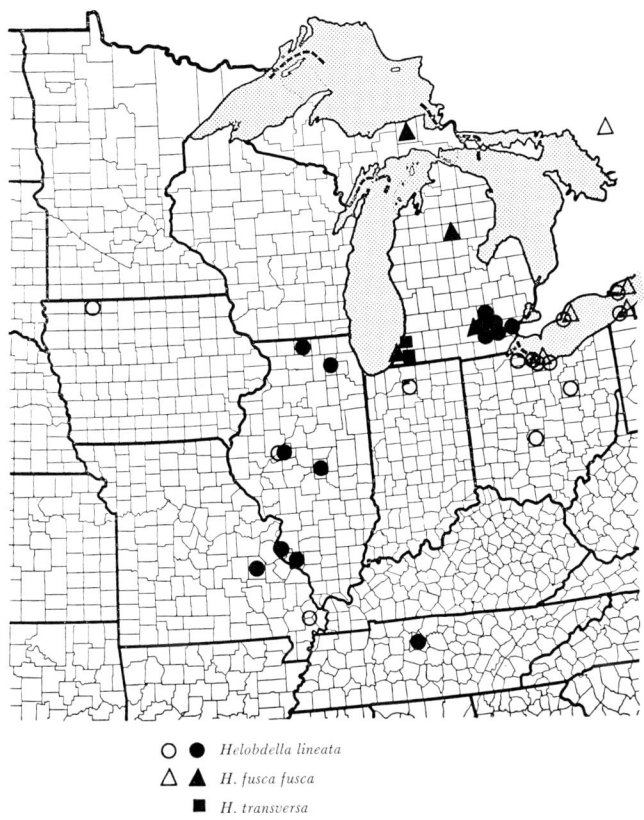

FIG. 27. *Helobdella lineata, H. fusca fusca, H. transversa.*

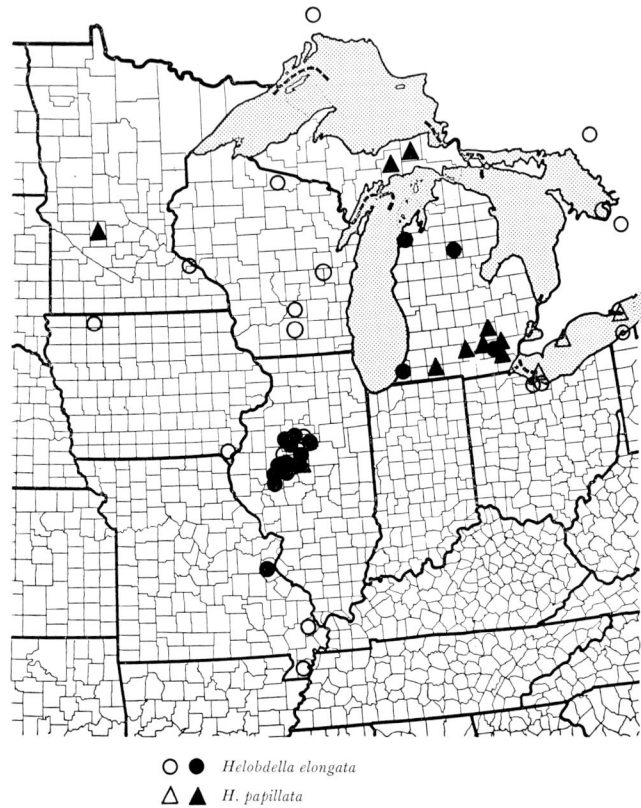

○ ● *Helobdella elongata*
△ ▲ *H. papillata*

FIG. 28. *Helobdella elongata, H. papillata.*

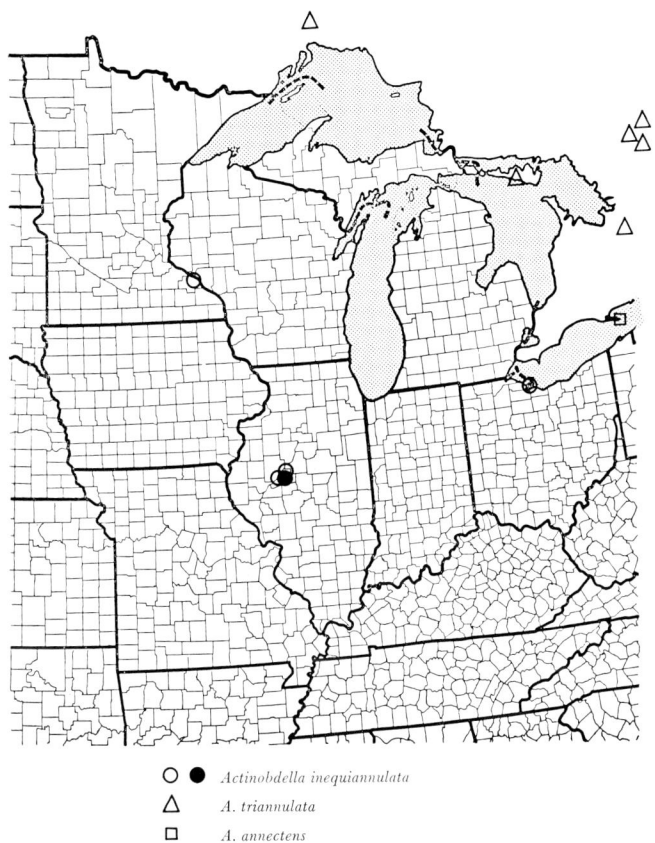

FIG. 29. *Actinobdella inequiannulata, A. triannulata, A. annectens.*

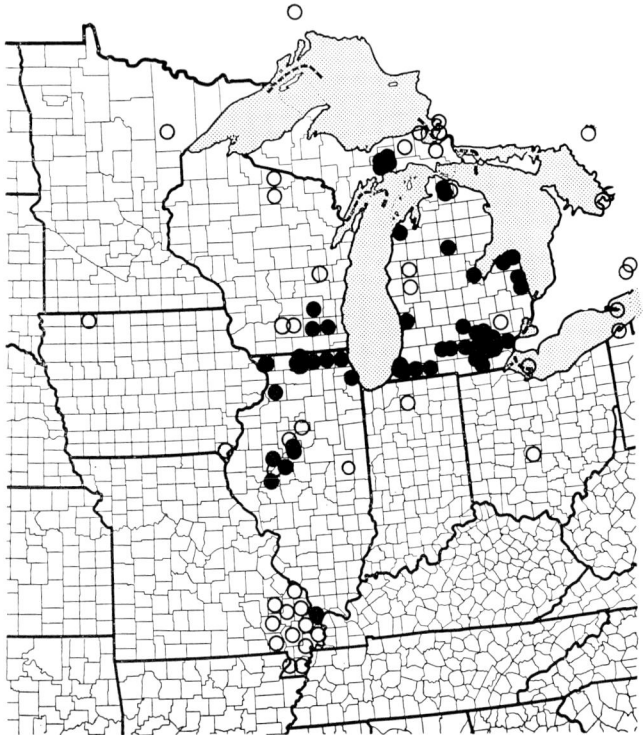

Fig. 30. *Erpobdella punctata*.

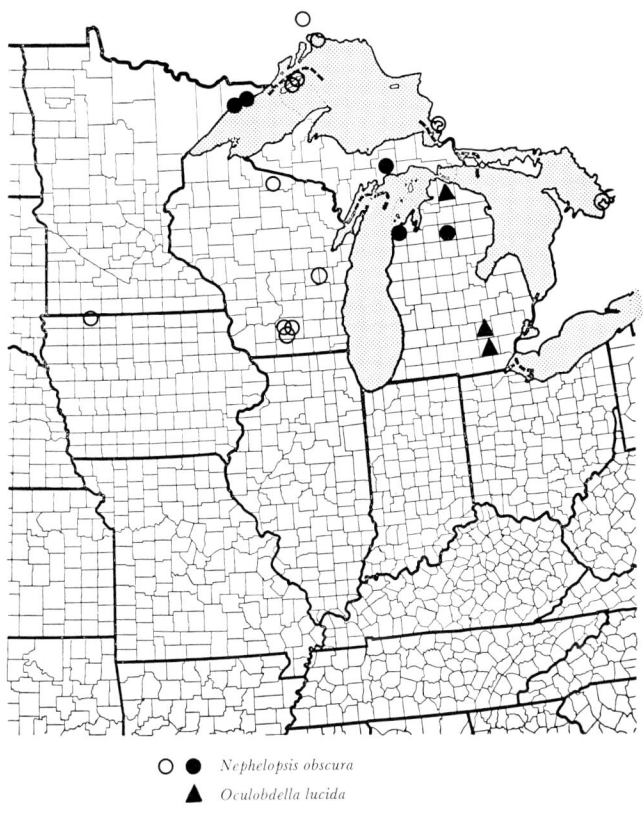

○ ● *Nephelopsis obscura*
▲ *Oculobdella lucida*

FIG. 31. *Nephelopsis obscura, Oculobdella lucida.*

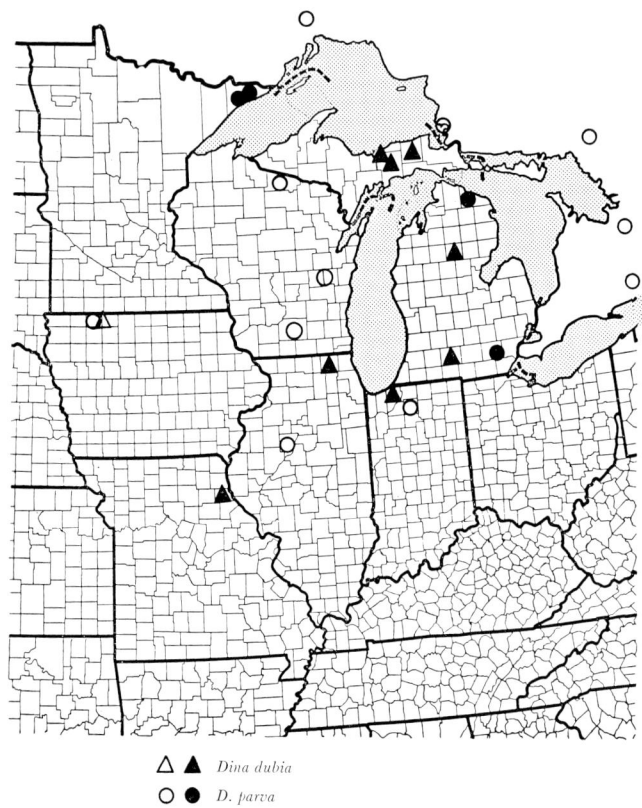

△ ▲ *Dina dubia*
○ ● *D. parva*

FIG. 32. *Dina dubia, D. parva*.

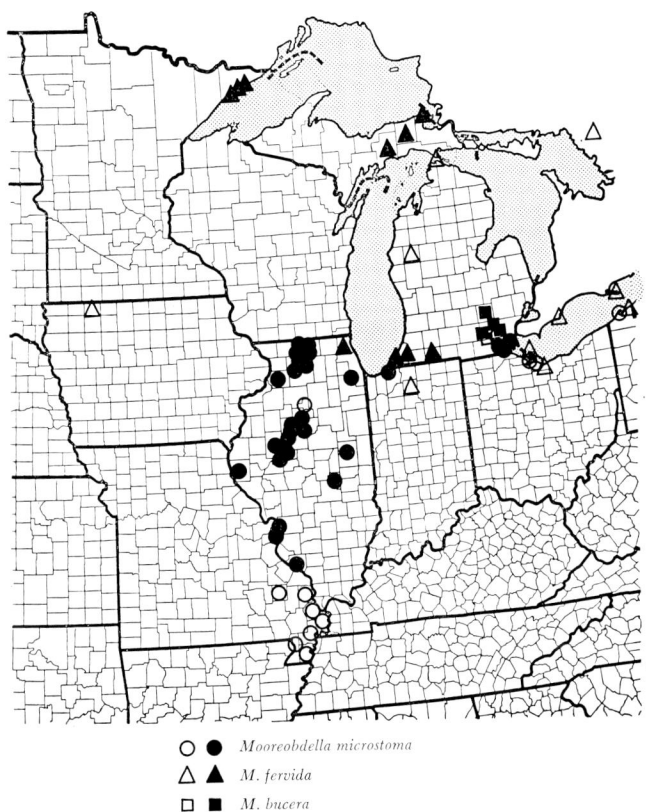

FIG. 33. *Mooreobdella microstoma, M. fervida, M. bucera.*

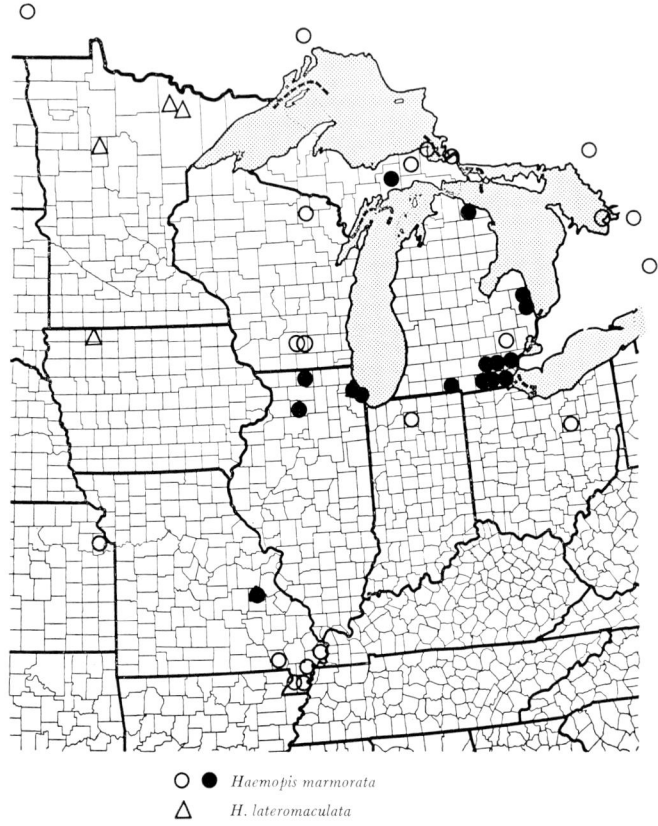

○ ● *Haemopis marmorata*
△ *H. lateromaculata*

FIG. 34. *Haemopis marmorata, H. lateromaculata.*

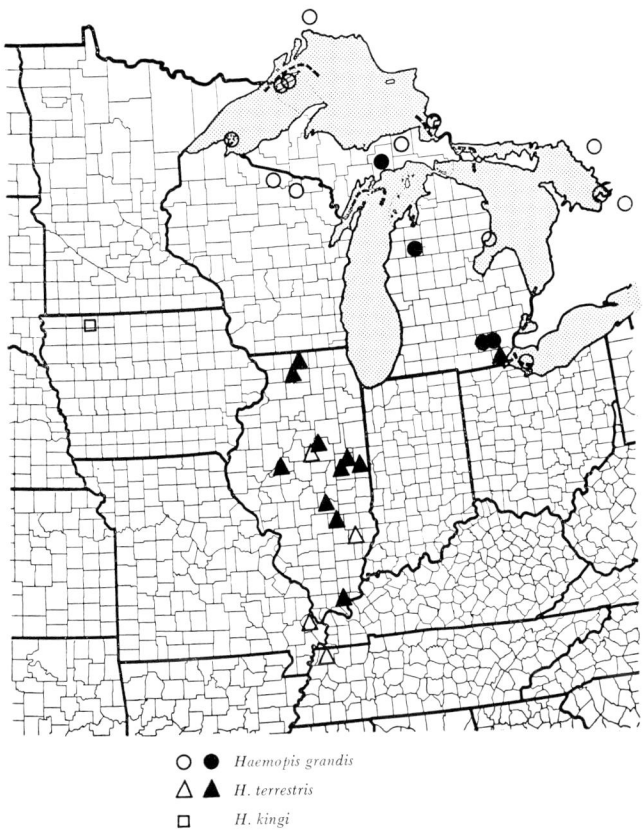

FIG. 35. *Haemopis grandis, H. terrestris, H. kingi.*

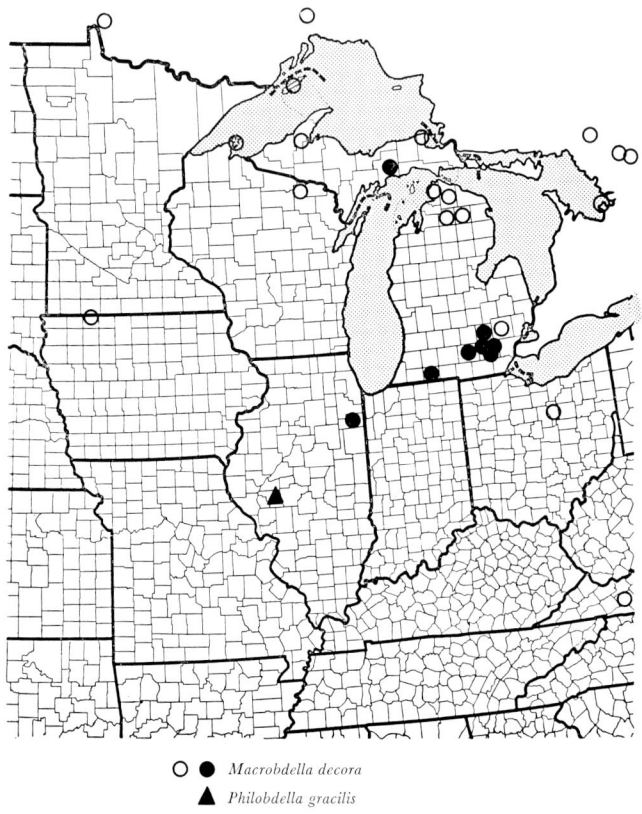

○ ● *Macrobdella decora*
▲ *Philobdella gracilis*

FIG. 36. *Macrobdella decora, Philobdella gracilis.*

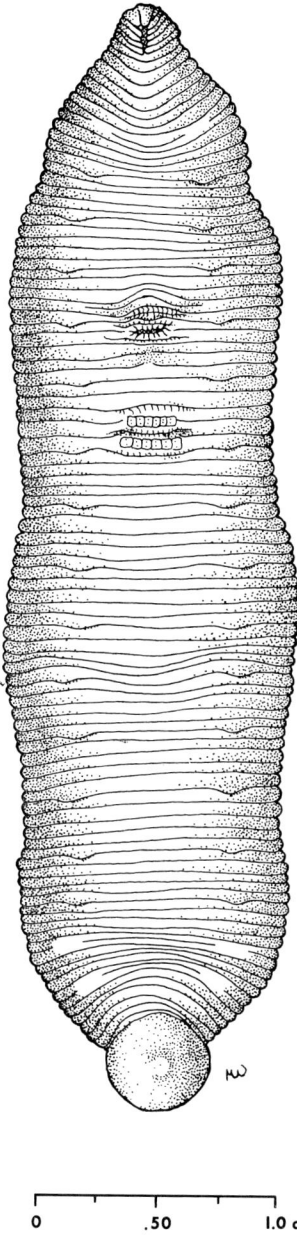

FIG. 37. *Macrobdella sestertia* (Cambridge, Massachusetts) (ventral view). Drawn by Mary Beth Welch, Duke University.

INDEX

Page numbers in italics refer to illustrations.

Accessory eyes. *See* Ocelli
Actinobdella, 28, 39, 73, 80
  A. annectens: description, 81; distribution, *139*
  A. inequiannulata: description, 40, 81, *109;* distribution, *139*
  A. triannulata: description, 81; distribution, 40, *139;* on fish, 40, 78
Agkistrodon, 72
Alligator: host for *Philobdella gracilis*, 72
Allolobophora: in crop of *Philobdella gracilis*, 72
Ambloplites, 22
Ambystoma, 11, 68
Americobdella valdiviana: from Chile, 65; earlier synonym of *Haemopis terrestris*, 65
Amphibians: *Batracobdella* in, 83; hosts for leeches, 79; leeches as vectors for blood parasites of, 73; leeches eating eggs of, 68, 72; *Macrobdella* in, 67-69; *Placobdella mon-* *tifera* in, 27. *See also* Frogs; Salamanders
Amphibious habits, 58. *See also* Terrestrial habits
Annulus, 77
Anoculobdella, 41
Aplodinotus, 28, 73
Aulastomum, 55

Barometric pressure: leeches responding to changes of, 1
Batracobdella: on amphibians, 81; description, 83, *110;* systematics, 9
  B. michiganensis: description, 13-14, 83, *110;* distribution, 13, *131*
  B. paludosa, 9
  B. phalera: description, 12, 83, *110;* distribution, 13, *131;* ecology, 13; food, 12-13; reproduction, 13; resemblance to *Actinobdella annectens*, 39; synonymy, 12
  B. picta: carrying *Helobdella stagnalis*, 31; description, 10, 83,

149

110; distribution, 12, *130;* ecology, 10-11; food, 11; in lymph spaces of bullfrog, 75; mated with *Placobdella parasitica,* 20; regulating population density of *Bufo americanus,* 11; reproduction, 11; synonymy, 10; vector for trypanosomes, 73
Bdellarogatis, 54
Birds: dispersal agent for leeches, 75; hosts for leeches, 79; *Macrobdella decora* in, 68; *Placobdella ornata* in, 22; *Theromyzon* in, 17. *See also Fulica; Marcea; Querquedula; Podilymbus*
Bloodsucking habits, 67, 79, 87. *See also* Medicine
Brooding, 78-79
Bufo americanus: host for *Batracobdella picta,* 11; host for *Macrobdella decora,* 68; natural populations regulated by *Batracobdella picta,* 11

Cannibalism: of adult leeches, 1, 58-59; of cocoons, 52-53
Carp. *See Cyprinus carpio*
Catostomus: host for *Actinobdella triannulata,* 40, 81
Caudal sucker: stalked in *Actinobdella,* 39-40, 81; stalked in *Placobdella pediculata,* 28, 40, 80
Chelydra serpentina as host for leeches: *Actinobdella annectens,* 81; *Philobdella gracilis,* 72; *Placobdella ornata,* 22; *Placobdella parasitica,* 20
Chitinous structures in leeches: cocoons in Erpobdellidae, *119;* dorsal plaque or scute in *Helobdella stagnalis,* 84, *113, 116;* teeth in Hirudinidae, 87
Chrysemys picta as host for leeches: *Placobdella hollensis,* 26; *P. ornata,* 22; *P. parasitica,* 20
Clemmys guttata, 20
Clepsine elegans: synonym for *Glossiphonia complanata,* 4
C. occidentalis: synonym for *Theromyzon meyeri,* 14-15
C. pallida: synonym for *Glossiphonia heteroclita,* 4
C. papillifera var. *b:* synonym for *Helobdella papillata,* 37
Cocoons: in Erpobdellidae, *119*
Coot. *See Fulica*
Cows: associated with leeches, 67, 70
Cyprinus carpio, 27
Cystobranchus, 85, *128*

Desmognathus: host for *Oligobdella biannulata,* 41, 83
Desiccation tolerances in *Placobdella parasitica,* 21
Dispersal agents, 75
Dina, 47, 86
D. anoculata: doubtful species, 47, 87
D. dubia: cocoon, *119;* description, 48, 87, *117, 118;* distribution, 48-49, 74, *142;* ecology, 48; enemies, 58; reproduction, *121;* synonymy, 48
D. lateralis: synonym for *Erpobdella punctata,* 47
D. parva: description, 47, 87, *117, 118, 127;* distribution, 48, 74, *142;* ecology, 48; synonymy, 47
Distribution maps, 3, *129-146*
Drumfish. *See Aplodinotus*

Ejaculatory duct, 86
Emydoidea blandingi: host for *Placobdella parasitica,* 20
Endemic genera of North America, 73
Endoparasitism: by *Batracobdella picta* in subcutaneous lymph spaces of frog, 11; by *Macrobdella ditetra* in teats of cow, 75; by *Nephelopsis obscura* in air bladder of fish, 75
Erpobdella, 43
E. octoculata: related to *E. punctata,* 43
E. punctata: cocoon, *119;* description, 43, 86, *117, 118;* distribution, 44-45, 74, *140;* ecology, 44; enemies, 7, 58; food, 44; life cycle, 44; migration, 44, 75; reproduction, 44; synonymy, 44; variations, *120*

INDEX   151

Erpobdellidae, 43-53, *117*
Eyes. *See* Ocelli

Figures, *109-147*
Fish as hosts for leeches, 78; *Actinobdella triannulata*, 40; *Macrobdella decora*, 68; *Nephelopsis obscura* in air bladder of, 46; *Piscicola salmositica*, 73; *Placobdella montifera*, 27; *P. ornata*, 22; *P. parasitica*, 20; *P. pediculata*, 28, 73. *See also Ambloplites; Aplodinotus; Catostomus; Cyprinus carpio; Ictalurus melas; Lepisosteus; Lepomis; Micropterus; Moxostoma; Salvelinus;* Sturgeon
Fossil leeches, 73. *See also* Chitinous structures in leeches; *Pontobdellopsis cometa*
Frogs as hosts for leeches: *Batracobdella picta*, 11; *Macrobdella decora*, 68; *M. ditetra*, 69; *Philobdella gracilis*, 72; *Placobdella parasitica*, 20. *See also Bufo americanus; Hyla; Rana*
Fulica, 17, 22

Ganglion, 78, 88
Glands: adhesive, 80; copulatory depressions, 87, *124;* copulatory, 66, 67, 69-71, 79, 87, *124, 147;* retractile papillae, *109*
Glossiphonia, 4-9
  G. complanata: brooding behavior, 7; description, 5, 79, *109, 127;* distribution, 7, 74, *129;* ecology, 6-7; food, 6-7, 31, 59; reproduction, 7, *114-115;* synonymy, 5; variations, 6
  G. complanata mollissima, 4
  G. heteroclita: description, 8, 80, *109, 127;* distribution, 9, 74, *130;* ecology, 8-9; reproduction, 9; synonymy, 8; variations, 8
  G. rudis: synonym for *Theromyzon rude*, 14
Glossiphoniidae, 4-42, 79
Glossiphoniinae: including *oculobdella*, 42; including *oligobdella*, 41
Gnathobdella, 79

Gonopore, 78, 86
Graptemys geographica: host for *Placobdella parasitica*, 20

Haementarinae, 41-42
Haementeria: synonym for *Placobdella*, 18
Haemopis: description, 87, *125;* recent revision, 54; systematics, 54. *See also Bdellarogatis; Mollibdella; Percymoorensis*
  H. grandis: carrying *Helobdella stagnalis*, 31; description, 60, 90, *123-125, 127;* distribution, 62, 74, *145;* ecology, 62; enemies, 7; synonymy, 55; variations, 56, *126*
  H. marmorata: carrying *Helobdella stagnalis*, 31; description, 56, 90, *123-125;* distribution, 59, 74, *144;* ecology, 58; enemies, 58-59; food, 58; mated with *Placobdella parasitica*, 20; migration, 58, 75; reproduction, 59; synonymy, 55; variations, 56, *126*
  H. kingi, 90-91, *145*
  H. lateromaculata, 90, *144*
  H. plumbea, 66, 90
  H. terrestris: description, 63, 79, 89, *123-125;* distribution, 65, *145;* ecology, 65; synonymy, 62-63; variations, *126*
Helisoma, 31, 36
Helobdella: description, 84, *112;* distribution, 29; systematics, 29-30
  H. elongata: description, 37, 84, *113;* distribution, 37, *138;* ecology, 37; synonymy, 37. *See also H. michaelseni*
  H. fusca: description, 36, 84, *112;* distribution, 36, *137;* polymorphism, 34-36; reproduction, 35; synonymy, 35, 37-38
  H. lineata: description, 34, 84, *112;* distribution, 35, 74, *137;* polymorphism, 34-35; synonymy, 33-35
  H. michaelseni, 37
  H. nepheloidea: synonym for *H. elongata*, 37

H. *papillata:* description, 38, 84, 112; distribution, 38, *138;* reproduction, 38; synonymy, 37-38

H. *punctatolineata:* from Puerto Rico, 30

H. *stagnalis:* carried by *Haemopis grandis,* 31; carried by *Haemopis marmorata,* 59; carried by *Macrobdella decora,* 31; description, 30, 84, *113, 127;* distribution, 33, 74, *136;* enemies, 58; reproduction, 31-33, *114-116;* synonymy, 30; variations, 31, 34

H. *transversa:* description, 38-39, 84, *113;* distribution, 38, *137*

H. *triserialis:* polymorphism, 34; from South America, 35

Hemoflagellates: spread by *Piscicola salmositica,* 73

Hirudiniasis, 75

Hirudinidae: etymology of, 54; systematic accounts of, 54-72, 79, 87-90, *123*

Hirudo medicinalis: not established in North America, 72, 87

Hyla, 11

Ictalurus melas: host for *Placobdella montifera,* 27

Illinobdella moorei, 73, 85, *128*

Insect larvae: as food for leeches, 27, 31, 37, 44, 46, 48, 50, 58, 62, 68

Key for leech identification, 2, 76-90

Kinosternon: host for *Philobdella gracilis,* 72

Leeching, 1, 67. See also Medicine

Lepisosteus: host for *Placobdella montifera,* 27

Lepomis: host for *Placobdella montifera,* 27

Lymnaea: not fed on by *Glossiphonia complanata,* 7; *G. heteroclita* harboring in mantle cavity of, 8

Macrobdella, 66, 73, 87
  M. decora: bloodsucking habits, 79; carrying *Helobdella stagnalis,* 31; description, 67, 87-88, *123-124;* distribution, 69, 74, 88, *146;* ecology, 67-68; enemies, 62; food, 67-68; life history, 67; reproduction, 68; synonymy, 66-67
  M. ditetra: description, 69, 88, *123-124;* distribution, 70, 74; ecology, 69-70; in teats of a cow, 75; synonymy, 69
  M. sestertia: description, 88, *147;* distribution, 75; synonymy, 70

Marcea, 22

Marvinmeyeria lucida. See *Oculobdella lucida*

Medicine: use of leeches in, 1, 67, 72

Menetus, 7

Metacercariae: in *Haemopis marmorata,* 59

Microbdella: synonym for *Oligobdella,* 40

Micropterus, 27

Migration of leeches, 1, 44, 58, 75

Mollibdella, 54, 60. See also *Haemopis*

Mooreobdella, 48, 73, 86
  M. bucera: cocoon, *119;* description, 51, 86, *117-118;* distribution, 53, 75, *143;* ecology, 53; population structure, 52, 53; reproduction, 52-53, *121;* synonymy, 51; variations, *120*
  M. fervida: cocoon, *119;* description, 49, 86, *117-118;* distribution, 50, 74, *143;* ecology, 50; synonymy, 49; variations, *120*
  M. microstoma: cocoon, *119;* description, 50, 86, *117-118;* distribution, 45, 51, 74, *143;* ecology, 50-51; synonymy, 50; variations, *120*

Moxostoma, 27

Mussels, 27

Natrix: host for *Philobdella gracilis,* 72

Nematomorphs: in *Erpobdella punctata,* 44

Nephelopsis, 45, 73
  N. obscura: cocoon, *119;* description, 44, 46, 79, 87, *117, 118;* distribution, 46-47, 74, *141;* ecology, 46; food, 46; in air bladder

of trout, 75; reproduction, 46; synonymy, 46-47; variations, *120*
Nephridiopore, *124*
Neurosecretion, 16
Notophthalmus, 11

Ocelli: accessory eyes of *Placobdella hollensis*, 25, 82, *127*; of leeches, 79, *127*
Oculobdella, 41
  O. lucida: description, 42, 83, *113*; distribution, 42, 74, *141*; ecology, 42; synonymy, 42
  O. socimulcensis, 41
Oligobdella, 40
  O. biannulata, 40-41, 75, 83

Parasites of leeches. See Hemoflagellates; Metacercariae; Nematomorphs; Trypanosomes
Percymoorensis, 54, 55, 63
Pharyngobdella, 79. See also Erpobdellidae
Philobdella: bloodsucking habits, 79; description, 87; distribution, 73; systematics, 70
  P. floridana: description, 88; distribution, 75; synonymy, 71
  P. gracilis: description, 71, 88, *123-124*; distribution, 72, 74, 146; ecology, 71-72; synonymy, 71
Physa, 6-7, 31
Pigments: green dissolves in ethanol, 20, 39
Piscicola, 85
  P. geometra, 85
  P. milneri, 85
  P. punctata, 73, 85, *128*
  P. salmositica: description, 85; food, 73; vector for blood parasites, 73
  P. virginica, 85
Piscicolaria, 73
  P. reducta, 85, *128*
Piscicolidae, 78, 84-85, *128*
Placobdella, 18, 81, 111
  P. hollensis: description, 25, 82, *111*, *127*; distribution, 26, *135*; food, 26; reproduction, 26; synonymy, 25

P. montifera: description, 27, 81, *111*; distribution, 27-28, *135*; food, 27; reproduction, 27; synonymy, 26-27
P. multilineata: description, 83; distribution, 23, 74; possible southern subspecies of *P. ornata*, 29; synonymy, 29. See also *P. ornata*
P. ornata: description, 22, 82-83, *111*; distribution, 23, 74, *133*; ecology, 22; enemies, 62; food, 22; reproduction, 23; synonymy, 21
P. papillifera: description, 24, 82, *111*; distribution, 24-25, *134*; food, 24; reproduction, 24; synonymy, 23
P. parasitica: description, 19, 82, *111*; desiccation, 21; distribution, 21, *132*; ecology, 20; mating with other species, 20; reproduction, 20; synonymy, 18-19
P. pediculata: description, 28, 80, 81, *111*; distribution, 28, *135*; food, 73, 78, 81; synonymy, 28
Podilymbus: host for *Theromyzon meyeri*, 17
Polymorphism in *Helobdella*, 29-30, 34-35, *112*
Pontobdellopsis cometa: possible leech fossil, 72
Pseudemys: host for *Placobdella ornata*, 22; host for *P. parasitica*, 20
Pulsatile vesicle, 85

Quaternary ice advance, 74
Querquedula: host for *Theromyzon meyeri*, 17

Rana: *catesbeiana* as host for *Batracobdella picta*, 11; for *Macrobdella decora*, 68; for *M. ditetra*, 69; for *Philobdella gracilis*, 72
  R. clamitans: host for *Philobdella gracilis*, 72
  R. grylio: host for *Philobdella gracilis*, 72
  R. pipiens: host for *Philobdella*

gracilis, 72; host for *Placobdella parasitica*, 20
Rhynchobdella, 78

Salamanders: hosts for *Batracobdella picta*, 11; hosts for *Macrobdella decora*, 68; hosts for *Oligobdella biannulata*, 41, 83. See also *Ambystoma*; *Desmognathus*; *Notophthalmus*; Trypanosomes
Salvelinus, 46, 68
Scute, dorsal. See Chitinous structures in leeches; *Helobdella stagnalis*
Semiscolex terrestris: synonym for *Haemopis terrestris*, 62
Snails: as hosts for *Glossiphonia complanata*, 6-7; for *Haemopis grandis*, 62; for *Haemopis marmorata*, 58; for *Helobdella fusca*, 36; not hosts for *Helobdella stagnalis*, 31; mantle cavity harboring *Glossiphonia heteroclita*, 8. See also *Helisoma*; *Lymnaea*; *Menetus*; *Physa*; *Stagnicola*
Snakes: as hosts for leeches, 72. See also *Agkistrodon*; *Natrix*
Snapping turtle. See *Chelydra serpentina*
Species: disjunct, 75; doubtful, 87; incipient, 75; introduced, 9, 15, 72, 85, 87; northern, 74; peripheral, 75; polymorphic, 29-30, 34-35, *112;* relict, 75; restricted geographically, 74-75; southern, 75; ubiquitous, 74
Stagnicola, 31
Sternothaerus as host for leeches: *Batracobdella phalera*, 12; *Placobdella ornata*, 22; *P. papillifera*, 24; *P. parasitica*, 20
Sturgeon, 68
Sucker. See *Catostomus;* Caudal sucker
Swimming in leeches, 18, 82
Synonymy, 2

Techniques: dissection, 77; preservation, 77
Teeth. See Chitinous structures in leeches
Terminology. See Annulus; Ejaculatory duct; Ganglion; Gonopore; Nephridiopore; Pulsatile vesicle; Trachelosome
Terrapene: host for *Placobdella ornata*, 22
Terrestrial habits, 63. See also Amphibious habits
Theromyzon: dispersed by birds, 75; systematics, 15-16
T. meyeri: description, 16-17, 80, *109, 127;* distribution, 18, *131;* ecology, 17; food, 17; reproduction, 17; synonymy, 16
T. rude: description, 80; neurosecretion, 16; type material, 15; synonymy, 15
T. tessulatum, 15
Toads. See Amphibians; Frogs
Trachelosome, 85
Trionyx: host for *Placobdella ornata*, 22
Trocheta, 46
Trypanosomes, 11
Turtles as hosts for leeches, 79, 81; *Actinobdella annectens*, 81; *Batracobdella phalera*, 12; *Macrobdella decora*, 67; *Philobdella gracilis*, 72; *Placobdella hollensis*, 26; *Placobdella ornata*, 22; *Placobdella papillifera*, 24; *Placobdella parasitica*, 20, 82. See also *Chelydra*; *Chrysemys*; *Clemmys*; *Emydoidea*; *Graptemys*; *Kinosternon*; *Pseudemys*; *Sternothaerus*; *Terrapene*; *Trionyx*

Wisconsinan ice advance, 74

Zoogeography of leeches, 73, 75. See also Distribution maps

About the Author

Roy T. Sawyer is assistant professor of biology at the College of Charleston, Charleston, South Carolina. He became interested in leeches as a student at the University of Michigan, where he earned his M.S. He received his Ph.D. from the University of Wales in Swansea, where he held a U.S. National Science Foundation predoctoral fellowship. Dr. Sawyer has written numerous other papers on marine as well as freshwater leeches; *North American Freshwater Leeches* is his first book.

University of Illinois Press